Operational Amplifier

Shrikrishna Pandurangji Yawale ·
Sangita Shrikrishna Yawale

Operational Amplifier

Theory and Experiments

Springer

Shrikrishna Pandurangji Yawale
Department of Physics and Electronics
Government Vidarbha Institute
of Science and Humanities
Amravati, Maharashtra, India

Sangita Shrikrishna Yawale
Pre Indian Administrative Services
Coaching Centre
Amravati, Maharashtra, India

ISBN 978-981-16-4187-9   ISBN 978-981-16-4185-5   (eBook)
https://doi.org/10.1007/978-981-16-4185-5

This Springer imprint is published by the registered company Springer Nature Singapore Pte Ltd.
The registered company address is: 152 Beach Road, #21-01/04 Gateway East, Singapore 189721,
Singapore

*Dedicated to my
father late Shri. Pandurangji Yawale
and elder brother late Shri.
Keshaorao Yawale*

Preface

Operational amplifier is the most widely used device in various electronics instruments. It is the building block of many analog circuits. Its versatility is proved by the important excellent characteristics and properties of the device. Being an important part of the instrumentation system, it has special weightage in design and fabrication of devices/circuits. Hence many electronics engineers prefer IC operational amplifier in electronic design. Therefore separate course is tailored in engineering and science curriculum. The first edition of this book is not rigour, but attempt is made to be systematic in approach. Authors feel that this book must meet the needs of students acquiring knowledge of operational amplifier. This book is more student-centric than rigorous treatments. This book *Operational Amplifier: Theory and Experiments* comprises two sections.

First section includes the basics of operational amplifier (Op-Amp) and its applications. Chapters 1–9 are devoted to the basic concepts. Various differential amplifiers, its classification, feedback in amplifier, parameters of operational amplifier, linear and nonlinear circuits, etc., are described in this book.

Chapter 1 includes the theory of differential amplifier, classification, AC/DC analysis, constant current sources, DC level shifter and output stage. Block diagram of operational amplifier is described. The single-stage architecture operational amplifier and its design are discussed in this chapter. The aim of this chapter is to provide the students with an overview of the basics of the operational amplifier and its design. The fundamental concept of differential amplifier is discussed to get the students and users well versed with the operational amplifier. The importance of electrical characteristics of Op-Amp is discussed in this chapter. Electronics engineers are always thinking about the compactness of the devices. Thus very large-scale integration (VLSI) design has the prime importance today. Operational amplifier is one of them having diversity of applications. It is a versatile device useful in many applications. This book is especially written for the students of electronics engineering and science which offers electronics specialization. The operational amplifier and applications is a concise and precise book in which the concept of operational amplifier is reviewed, since operational amplifier is a multitasking device used in many medical instruments for application of very low intensity signals obtaining prominent place in electronic instruments such as ECG, EEG, etc. This book comprises two sections, which covers almost all aspects of the operational amplifier.

The concept of feedback in amplifiers is equally important. Without feedback design, amplifier is worthless. So as to get better understanding of both types of feedbacks, positive and negative, the theory of feedback in amplifier is discussed in Chap. 2. Introduction, types of negative feedback, non-inverting amplifier, i.e. voltage series feedback and effect of feedback on various important properties, viz. closed loop voltage gain, input impedance, output impedance, frequency response, etc., are reported. The voltage follower circuit is also discussed. Similarly the concept of voltage shunt feedback in amplifier, i.e. inverting amplifier and its effect, on the above-discussed properties is reported in this chapter.

Chapter 3 treats the parameters of operational amplifier and its applications as AC/DC amplifier. While studying the Op-Amp, students should know the various parameters of Op-Amp such as input bias current, output offset current, input offset voltage, open loop gain, CMRR, input capacitance, slew rate, SVRR and offset voltage adjustment. Inverting and non-inverting AC/DC amplifiers and instrumentation amplifier are important for consumer as well as industrial applications. Students and instructors should be aware of the knowledge of designing the instrumentation amplifier and its applicability. Differential and three Op-Amp instrumentation amplifiers are discussed in this chapter. Applications of Op-Amp as linear devices as current to voltage and voltage to current converter, voltage and current measurements, summings, scaling and subtractor, integrator/differentiator circuits and analog computation, etc., are included in Chap. 4. The data acquisition and signal processing are the main components of any instrumentation system. Both circuits employed analog to digital (A/D) and digital to analog (D/A) conversion. Such converter circuits utilize Op-Amp. The importance of A/D and D/A converters and various types of them are discussed. As we have discussed that Op-Amp is a multitasking and multidimensional versatile device, it can be used in nonlinear functional circuits like mixers, modulators, rectifiers, etc. The important aspect in many devices is precision rectifier and peak detector that have been discussed. Chapter 5 contains logarithmic and exponential amplifiers, multiplier and amplitude modulator and so on. In a data acquisition system sample and hold circuit have prime importance. The concept of this circuit is also discussed. The circuits for wave generation of sinusoidal, square and triangular are described in Chap. 6. Classifications of oscillators such as phase shift, Wein Bridge, Colpitts and Hartley oscillator and non-sinusoidal oscillators such as astable and monostable multivibrators, their working, function generator, comparator and Schmitt trigger are described. More attention is paid to the pedagogy, explanation and working of circuits. Active filters and phase-locked loop (PLL) and its applications are discussed in Chap. 7. To get acquainted with the design of active filters and the applicability in instrumentation, low-pass, high-pass and band-pass filters are explained. Classification and the working of Butterworth first- and second-order filters are described. Phase-locked loop and operating principle have been used in many applications such as frequency shift keying, decoders, frequency multiplier and translators. The application of PLL as frequency multiplier and translator is described in this chapter.

Chapter 8 contains frequency-dependent negative resistance and gyrator. The gyrator is an important element in integrated circuits which replace the physical inductor in the circuit. The real inductors cannot be fabricated in the ICs, but this could be possible with gyrators. Gyrator is a simulated circuit of inductor. Large value inductors and extensive adjustable inductance range are easily implemented by using special electronic circuit having combination of resistors, capacitors and operational amplifiers, such a circuit is called gyrator. It is a two-port device or network element called hypothetical fifth linear lossless passive element after inductor, capacitor, register and ideal transformer. The theory of gyrator and its working are described in this chapter. Replacement of real iron and air-core inductors by gyrator is discussed. The advantages and disadvantages are also reported. During old; days before invention of gyrator it was an unbelievable thing that inductor without turns of wire on any core is not possible. Similarly frequency-dependent negative resistance (FDNR) or D-element is an active element that exhibits real negative resistance views like unusual capacitor. This can be fabricated using resistors, capacitors and operational amplifiers. LC low-pass and high-pass filters can be transformed using FDNR. In a transformation capacitors are replaced by FDNR inductance by resistance and resistance by capacitor. Gyrator is called synthetic inductor also. It is a negative impedance inverter whose input impedance is proportional to negative of the load admittance.

Chapter 9 describes noise in operational amplifier. Physically it is the differential amplifier; therefore the noise associated with differential amplifier is the noise generated in operational amplifier. While designing the circuit of any amplifier, the noise generated due to various components in amplifier circuit cannot be neglected especially when weak signal amplification is required. The noise in input signal is not the only source of noise, but fluctuations in a supply voltage or current, loose connection, Brownian motion of electrons in conductor and the transportation of the charges near the junction of semiconductor, etc., are the factors responsible for the generation of electric noise. Thus while designing the amplifier, the electronics engineer or designer has to take into account the various noise sources and how to suppress them. There are various types of noise, viz. Johnson, Schottky, low frequency, popcorn, thermal, shot, etc. In this chapter the classification of noise and their sources are reported. Generally the classification is made as interference and inherent noises. Theory of various types of noise and the noise generated in resistors and capacitors are discussed. Similarly operational amplifier noise model and its analysis in inverting and non-inverting mode have been reported. The equivalent noise models and calculation of total noise from them along with noise figure are given in this chapter.

Second section of the book covers the experiments of operational amplifier. Nearly about 17 experiments of Op-Amp are described. University students are well aware about the theory of Op-Amp, but how to perform the experiment is really a difficult task for them. Thus laboratory exercise in terms of experiments is described in the book. Before starting the laboratory work every laboratory exercise requires power supply; hence design of dual polarity $\pm$ 15 volt power supply is also discussed. Relevant bibliography is given at the end of this book. This book is useful for the students of Bachelor of Engineering, master's degree in Science (Physics and Electronics) and Diploma in Engineering. Our teachers, professors and students encouraged us to write the book on operational amplifier. Actually it was in our mind long back but could not complete early due to our busy schedule. The COVID-19 pandemic provided us an opportunity in lockdown to write the book due to closure of our institute. Our research students and our elder daughter Deepika inspired us to write the book. The excellent opportunity of work from home is encashed by writing this book. We are indebted to Director of Government Vidarbha Institute of Science and Humanities, Amravati (India), for giving permission to write this book. We also gratefully acknowledge our daughters, Deepika and Sanika, for their help in checking manuscript and giving valuable suggestions during the completion of this book. This book aimed to serve as textbook for the undergraduate and postgraduate students of electronics as well as physics in Indian universities. In addition to this more emphasis is given on the theoretical aspects along with experiments. The chapters written in this book are helpful for the students and teachers. This book as such cannot be a one-man project although it is impossible to write this book without timely help of faculty colleagues and our research/post graduate students who have contributed a lot. We express gratitude to them for their comments, suggestions and corrections. Specific thanks to our mentor and guide, Ex-Professor Dr. Chandrashekhar Adgaonkar, Head of the Department of Electronics, Institute of Science, Nagpur, who has inspired us to write this book and gave valuable suggestions. I place on record my gratitude to my research students, Principal Dr. Yogendra Gandole, Prof. Dr. Deepak Dhote, Dr. V. K. Sewane, Prof. S. B. Sawarkar, Dr. S. M. Palhade, Dr. Pritesh Jadhao, Mr. Kamlesh Banarase, Dr. Ramesh Zade, Mr. Pranav Awandkar and Dr. Dhananjay Bijwe for their constant support and encouragement during the preparation of this book. Especially I acknowledge my sincere thanks to Dr. Bhupesh Mude and Dr. Kushal Mude for taking the efforts to handover the handwritten manuscript to the typist during lockdown situation in Mumbai during pandemic.

I am very much thankful to Mr. Anil Devlekar, Mumbai, for typing the whole manuscript and the diagrams in the respective chapters of the book.

Finally, I express my sincere thanks to the publisher, Springer Publications Pvt. Ltd., for meticulous processing of the manuscript during various stages of publication of this book.

Amravati, India   Shrikrishna Yawale
June 2021   Sangita Yawale

Prologue

Today technology has evolved to bring the world closer. Every aspect of the daily life cannot be complete without using the electronic devices. Most of the people are using electronic gadgets right from kitchen appliances to smartphones. In such a world called electronics world the compact or functional devices play a very important role. One of the devices is the operational amplifier (Op-Amp) more popularly called as Op-Amp which is a versatile device utilized in many applications. Looking at the properties of the Op-Amp, it is a multitasking versatile device. Its tremendous amount of gain and differential input make it simpler and easy to design many types of electronic circuits. This is an unparalleled device as compared to the properties.

Operational amplifier can be used for various applications. Its input/output impedance and large bandwidth and gain make their implementation most basic in electronic circuits. Op-Amps are quite easy to use in linear as well as in nonlinear device.

In analog circuit design it is unmatched with other devices. It may be called the basic building block of analog electronic circuits. Its capacity to operate with positive as well as negative feedback in non-sinusoidal waveform generation is very important. The precision rectifier is also equally important for low-voltage rectification. We can use it in any manner for fabrication and design of analog circuits. Not only this but mathematical operations such as addition, subtraction, multiplication, division, logarithm and antilogarithm, and solutions to differential equations are the rare properties found in this amplifier. The unique properties of Op-Amp can pave the new ways to build the good electronic circuits.

The Op-Amps like LM 741, CA 3130, TL 071, LM 311, CA 3140 and LM 358, etc., have excellent performance. The IC 741 is a very popular and low-price Op-Amp available in any electronics market easily. This device can be made compatible with digital ICs also. This is a widely used electronic device in consumer, industrial and scientific electronic circuits. Instrumentation amplifier is a good example where Op-Amp is used for amplification of low intensity signal to desired level. While designing any circuit, very few external components like capacitors and resistors are required.

Observing to the important properties of this device it has been decided to write about the theory of the operational amplifier and its applications. This book can be used as text as well as reference book for engineering and science graduate students. The objective of this book is to describe the various applications of operational amplifier along with theory.

June 2021   Shrikrishna Yawale
                                                                    Sangita Yawale

Contents

About the Authors

Dr. Shrikrishna Yawale is from the Department of Physics and Electronics, Government Vidarbha Institute of Science and Humanities. He did his post-graduation in Physics specializing in electronics. He is continuously engaged in teaching the subject of electronics for the last 34 years. He has published over 100 research papers in electronics and instrumentation in reputed national and international journals. Dr. Yawale successfully guided several students for the Ph.D. degree in electronics. He is a member of the Board of Studies in Electronics of Sant Gadge Baba Amravati University. His areas of research are solid-state physics, semiconducting glasses, gas sensors, conducting polymers, and instrumentation. He is the co-author of the book on electronics: Advanced Digital Techniques, Microprocessors, and 8051 microcontrollers.

Dr. Sangita Yawale is from Pre Indian Administrative Services Coaching Centre, Amravati. She did her post-graduation in Physics specializing in electronics. She has 24 years of teaching and 10 years of administrative experience. She has been continuously engaged in research and has published over 100 research papers in national and international journals. She has also attended conferences and delivered lectures. She has successfully guided several students for their Ph.D. degrees. Her research areas are materials science and gas sensor.

Chapter 1
Differential Amplifier

This chapter includes the theory of *differential amplifier*, classification, AC/DC analysis, constant current sources, DC level shifter and output stage. The differential amplifier is a backbone of the operational amplifier (Op-Amp) rather we may call it as a basic building block. The block diagram of operational amplifier is described. The single-stage architecture operational amplifier and its design are discussed in this chapter. Aim of this chapter is to provide the students an overview of the basics of the operational amplifier and its design. The fundamental concept of differential amplifier is discussed to get the students and users well versed with the operational amplifier. Importance of electrical characteristics of Op-Amp is discussed in this chapter. Electronics engineers are always thinking about the compactness of the devices. Thus, very large-scale integration (VLSI) design has the prime importance today. Operational amplifier is one of them having diversity of applications. It is a versatile device useful in many applications. Since operational amplifier is a multitasking device used in many medical instruments for the application of very low intensity signals obtain prominent place in electronic instruments.

1.1 Introduction

The frequency response of the RC-coupled amplifiers is limited due to coupling condensers. To achieve the large gain of amplifier, a number of common emitter amplifiers are cascaded through resistance capacitor (RC) coupling. In such amplifiers, gain will increase but the frequency response of the amplifier limits at lower and higher frequencies and the gain falls. The basic reason is that the reactance of the coupling capacitors will change. Also at higher frequencies, stray and *Miller capacitors* become appreciable. Therefore, the bandwidth of RC-coupled amplifier limits. For larger bandwidth and higher gain (i.e. ideally infinite bandwidth and infinite gain, theoretically) if coupling capacitors are removed and the

© The Author(s), under exclusive license to Springer Nature Singapore Pte Ltd. 2022
S. Yawale and S. Yawale, *Operational Amplifier*,
https://doi.org/10.1007/978-981-16-4185-5_1

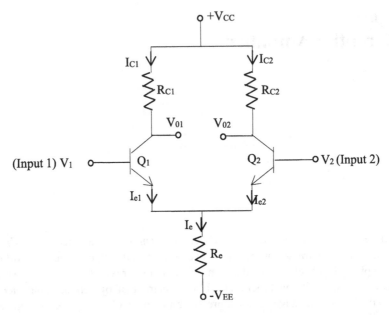

Fig. 1.1 Differential amplifier

signal source is directly connected, then the bandwidth gets improved, and such class of amplifiers is called DC amplifier.

However, these amplifiers have some drawbacks such as shift in the operating point, change in thermal stability basing. To get rid of all these drawbacks of DC amplifier, a class of differential amplifier is invented which has larger gain, larger bandwidth, large input impedance, low output impedance, more stability and zero drift, etc.

The differential amplifier is a modification of DC amplifier. It is a device used to amplify the difference between two input voltages and suppress any voltage common to the two inputs. These amplifiers are also called emitter-coupled amplifiers as two transistors emitters are coupled together and connected through common emitter resistance to negative terminal of dual power supply. The emitters of two common emitter amplifiers are coupled and connected to an emitter resistance as shown in Fig. 1.1.

1.2 Classification of Differential Amplifier

Generally, differential amplifiers or emitter-coupled amplifiers are classified as follows:

(a) Dual input balanced (dual) output
(b) Dual input unbalanced (single ended) output

(c) Single input balanced (dual) output
(d) Single input unbalanced (single ended) output.

The classification is made on the basis of two inputs and two outputs.

The first stage of operational amplifier is high gain differential amplifier rather it is dual input balanced output differential amplifier. Hence, we will discuss only dual input balanced output differential amplifier.

Figure 1.1 shows the basic diagram of dual input balanced output differential amplifier.

The differential amplifier is capable of amplifying DC as well as AC signals. These amplifiers are widely used in instrumentation system for comparing two input signals. In the circuit, Q_1 and Q_2 are identical transistors, and collector resistances R_{C1} and R_{C2} have equal value. Emitter resistance is common for both transistors Q_1 and Q_2. Both transistors are fabricated in a single package of IC so that they become identical. The input signal is applied through signal sources V_1 and V_2 having very small source resistances r_{s1} and r_{s2}. All voltages are measured with respect to ground.

1.3 DC Analysis

In DC analysis we have to determine operating points I_{CQ} and V_{CEQ} of differential amplifier for that it is necessary to draw DC equivalent circuit. In earlier Fig. 1.1, the input signal sources are not shown while deriving the values of collector current I_{CQ} and collector to emitter voltage V_{CEQ} (DC values) at operating point of transistors, the input sources connected at two inputs of the amplifier must be grounded in equivalent circuit, but the internal resistances r_{s1} and r_{s2} need to be considered. Figure 1.2 shows DC equivalent circuit of emitter-coupled or differential amplifier. In a differential amplifier, it is necessary to have both transistors Q_1 and Q_2 matched, i.e. $Q_1 = Q_2 = Q$ and collector resistors $R_{C1} + = R_{C2} = R_C$.

Since the differential amplifier is symmetrical in all aspects, the operating point values of I_{CQ} and V_{CEQ} are identical for both transistors Q_1 and Q_2. Let us determine the operating point values I_{CQ} and V_{CEQ}.

The Kirchhoff's voltage law expression for base–emitter junction yields— (transistor Q_1)

$$r_s I_B + V_{BE} + 2 I_e R_e - V_{EE} = 0$$
$$\text{or} \quad r_s I_B + V_{BE} + 2 I_e R_e = V_{EE} \tag{1.1}$$

However the DC current gain $\beta_{DC} = \frac{I_c}{I_B} = \frac{I_e}{I_B}$ and $I_c \approx I_e$. Thus, the emitter current flowing through transistor Q_1 can be determined from Eq. (1.1) by substituting the value of $I_B = \frac{I_e}{\beta_{DC}}$.

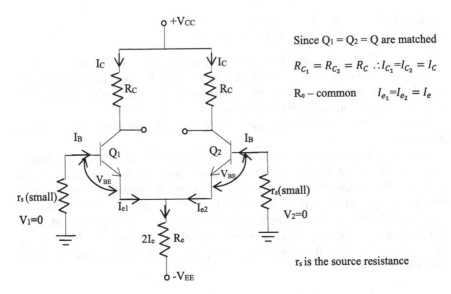

Since $Q_1 = Q_2 = Q$ are matched

$R_{C_1} = R_{C_2} = R_C \therefore I_{C_1} = I_{C_2} = I_C$

R_e – common $I_{e_1} = I_{e_2} = I_e$

r_s is the source resistance

Fig. 1.2 DC equivalent circuit of differential amplifier

$$\therefore 2I_e R_e = V_{EE} - V_{BE} - r_s \frac{I_e}{\beta_{DC}}$$

$$I_e \left(2R_e + \frac{r_s}{\beta_{DC}} \right) = V_{EE} - V_{BE}$$

$$\text{or} \quad I_e = \frac{V_{EE} - V_{BE}}{\left(2R_e + \frac{r_s}{\beta_{DC}} \right)}$$

Since r_s is the internal resistance of the sources, it is very small, and the DC current gain of the transistor is large (i.e. $\beta_{DC} \approx 100$), then

$$\frac{r_s}{\beta_{DC}} \ll 2R_e$$

$$\therefore \quad I_e = \frac{V_{EE} - V_{BE}}{2R_e} \tag{1.2}$$

$$I_{CQ} = \frac{V_{EE} - V_{BE}}{2R_e} \quad (\because I_e \approx I_c)$$

Equation (1.2) indicates that emitter current I_e is independent of collector resistance R_C, and it will be decided by V_{EE} value only.

Let us determine the collector to emitter voltage V_{CE}. The voltage at the collector V_C is given by

$$V_C = V_{CC} - I_C R_C \tag{1.3}$$

Since source resistance r_s is small, the voltage drop across it is small. Thus

$$V_{CE} = V_C - V_e \tag{1.4}$$

($\because$ emitter voltage at Q_1 is equal to $-V_{BE}$ is small).
Substituting value of V_C from Eq. (1.3) in Eq. (1.4), we get

$$\begin{aligned} V_{CE} &= (V_{CC} - I_C R_C) - (-V_{BE}) \\ V_{CEQ} &= V_{CC} - I_C R_C + V_{BE} \end{aligned} \tag{1.5}$$

Equations (1.2) and (1.5) are called DC analysis equations which are applicable to all classified differential amplifiers. Thus from these equations we can determine operating point values.

1.4 AC Analysis

By using AC analysis we can obtain expression for voltage gain A_d and input resistance R_i of the differential amplifier. The small signal AC equivalent r-parameter circuit is shown in Fig. 1.3. For small signal r-equivalent model set-up DC voltages, $+V_{CC}$ and $-V_{EE}$ in the circuit equal to zero. The r-parameter treatment is simple and easy as compared to others. Therefore this treatment is used for the derivation of various parameters. The AC equivalent of common emitter (CE) configuration is used. In the circuit $I_{e1} = I_{e2}$, therefore $r_{e1} = r_{e2} = r_e$, where r_{e1} and r_{e2} are the AC emitter resistance of transistors Q_1 and Q_2, respectively. Both transistors are replaced with their equivalent circuit.

Let us determine AC parameters of the differential amplifier for small signal AC equivalent r-model.

(a) *Voltage gain A_d:*

Because of common emitter configuration, the voltage across each collector resistor is out of phase by $180°$ with inputs V_1 and V_2.

Apply the Kirchhoff's voltage law to loop I and II in Fig. 1.3, we get

$$V_1 - r_{s1} i_{b1} - r_e i_{e1} - R_e(i_{e1} + i_{e2}) = 0 \tag{1.6}$$

$$V_2 - r_{s2} i_{b2} - r_e i_{e2} - R_e(i_{e1} + i_{e2}) = 0 \tag{1.7}$$

The AC current gain of the transistor is given by

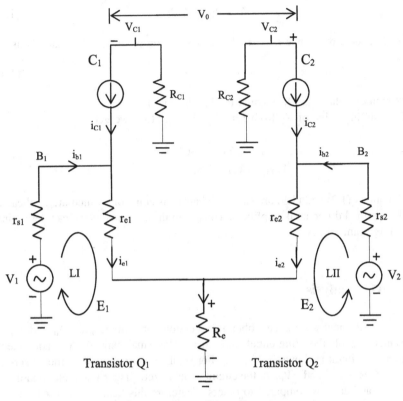

In the circuit - $R_{C_1} = R_{C_2} = R_c$, $r_{e_1} = r_{e_2} = r_e$

Fig. 1.3 Small signal AC equivalent r-model

$$\beta_{ac} = \frac{i_{e1}}{i_{b1}} \text{ and } \beta_{ac} = \frac{i_{e2}}{i_{b2}}$$

Substituting values of i_{b1} and i_{b2} in Eqs. (1.6) and (1.7), we get

$$V_1 - r_{s1}\frac{i_{e1}}{\beta_{ac}} - r_e i_{e1} - R_e(i_{e1} + i_{e2}) = 0 \qquad (1.8)$$

Similarly

$$V_2 - r_{s2}\frac{i_{e2}}{\beta_{ac}} - r_e i_{e2} - R_e(i_{e1} + i_{e2}) = 0 \qquad (1.9)$$

Generally $\frac{r_{s1}}{\beta_{ac}}$ and $\frac{r_{s2}}{\beta_{ac}}$ values are very small as compared to R_e and r_e, therefore we can neglect them. So Eqs. (1.8) and (1.9) will be

$$V_1 = r_e i_{e1} + R_e i_{e1} + R_e i_{e2}$$

and

$$V_2 = r_e i_{e2} - R_e i_{e1} + R_e i_{e2}$$

or

$$V_1 = (r_e + R_e)i_{e1} + R_e i_{e2} \tag{1.10}$$

and

$$V_2 = R_e i_{e1} + (r_e + R_e)i_{e2} \tag{1.11}$$

On solving Eqs. (1.10) and (1.11) simultaneously by using usual method, we get i_{e1} and i_{e2} values.

$$i_{e1} = \frac{(r_e + R_e)V_1 - R_e V_2}{(r_e + R_e)^2 - R_e^2} \tag{1.12}$$

$$i_{e2} = \frac{(r_e + R_e)V_2 - R_e V_1}{(r_e + R_e)^2 - R_e^2} \tag{1.13}$$

The output voltage V_0 is given by

$$V_0 = V_{C_1} - V_{C_2}$$
$$= -R_C i_{c_2} - (-R_C i_{c_1})$$
$$= R_C(i_{c_1} - i_{c_2})$$

$$\text{or} \quad V_0 = R_C(i_{e_1} - i_{e_2}) \quad (\because i_c \approx i_e) \tag{1.14}$$

Substituting i_{e_1} and i_{e_2} value in Eq. (1.14), we get,

$$V_0 = R_C \left[\frac{(r_e + R_e)V_1 - R_e V_2}{(r_e + R_e)^2 - R_e^2} - \frac{(r_e + R_e)V_2 - R_e V_1}{(r_e + R_e)^2 - R_e^2} \right]$$

On solving this equation, we get

$$V_0 = \frac{R_C}{r_e}(V_1 - V_2)$$
$$= \frac{R_C}{r_e}V_d \quad (\because V_d = V_1 - V_2) \tag{1.15}$$

Thus voltage gain

$$A_d = \frac{V_0}{V_d} = \frac{R_C}{r_e} \tag{1.16}$$

From Eq. (1.16), we can say that the voltage gain is independent of R_e, the emitter resistance.

(b) *Input resistance R_i*

Differential input resistance is defined as the equivalent resistance between either of input terminals and ground. Thus to find R_{i1}, the V_2 input should be grounded, and to find R_{i2}, V_1 input should be grounded.

$$\therefore \quad R_{i_1} = \frac{V_1}{i_{b_1}} \quad \text{when } V_2 = 0$$

$$= \frac{V_1}{i_{e_1}/\beta_{ac}} = \frac{V_1 \beta_{ac}}{i_{e_1}} \tag{1.17}$$

Substituting value of i_{e_1} from Eq. (1.12) in (1.17), we get

$$R_{i_1} = \frac{\beta_{ac} V_1}{\frac{(r_e + R_e)V_1 - R_e V_2}{(r_e + R_e)^2 - R_e^2}}$$

On solving above equation, we get

$$R_{i_1} = \frac{\beta_{ac} r_e (r_e + 2R_e)}{(r_e + R_e)}$$

But emitter resistance $R_e \gg r_e$, hence $(r_e + R_e) \approx R_e$ and $(r_e + 2R_e) \approx 2R_e$

$$\therefore \quad R_{i_1} = \frac{2\beta_{ac} r_e R_e}{R_e} \tag{1.18}$$

$$R_{i_1} = 2r_e \beta_{ac}$$

Similarly, we can calculate the value of R_{i_2} by substituting i_{e_2} from Eq. (1.13) in (1.17), we get

$$R_{i_2} = \frac{V_2}{i_{b_2}} = \frac{V_2 \beta_{ac}}{i_{e_2}} \quad \text{when } V_1 = 0 \tag{1.19}$$

$$\therefore \quad R_{i_2} = 2r_e \beta_{ac}$$

From Eqs. (1.18) and (1.19), we can see that input resistance of differential amplifier depends on AC urrent gain β_{ac} of the transistor and AC-emitter resistance r_e. Also the input resistance of differential amplifier on either side is equal.

(iii) Output resistance R_0

It is the effective resistance measured between output terminal and ground. It means this resistance is the collector resistance R_C (see Fig. 1.3)

$$\therefore \quad R_{01} = R_{02} = R_C \tag{1.20}$$

(iv) Common mode rejection ratio (CMRR)

In the circuit diagram (Fig. 1.3) when the same signal having equal magnitude is applied to both inputs, the expected output should be zero. The capability of the amplifier to reject the common mode signal is indicated in terms of CMRR. Thus CMRR is the ratio of differential mode gain A_d to common mode gain A_c of the amplifier.

$$\therefore \quad \text{CMRR} = \frac{A_d}{A_c} \tag{1.21}$$

The CMRR decides the quality of the difference amplifier, higher the value of CMRR better would be the amplifier.

(e) Differential mode and common mode gain

In a differential amplifier, amplified voltage is related to the difference $(V_1 \sim V_2)$ of the inputs. The voltage in such a case is called differential voltage gain A_d.

$$\therefore \quad A_d = \left| \frac{\text{output}}{\text{Differential mode input}} \right| = \frac{V_0}{(V_1 \sim V_2)} \tag{1.22}$$

If inputs of differential amplifier are $V_1 = V_2$, then output should be zero (ideal). But practically we get certain output voltage. The gain of the amplifier in such a case is called common mode gain A_c.

$$\therefore \quad A_c = \left| \frac{\text{output}}{\text{Common mode input}} \right| = \frac{V_0}{C_{mi}} \tag{1.23}$$

A good quality differential amplifier possesses smaller common mode gain A_c. Ideally it should be zero. The A_c can be reduced to a very small value by proper selection of high-current gain β value transistors and higher value of emitter resistance R_e. The high β value of transistor can be attained by fabricating *Darlington pair* in place of each transistor. Another way is to increase the value of emitter resistance R_e, but higher value of R_e leads to the requirement of higher battery supply. The remedy for larger value of R_e is to replace R_e by constant current source which fulfils the condition of infinite R_e. In such case, a smaller battery supply is required.

1.5   Constant Current Source or Current Bias

It is a circuit designed for constant current supply independent of load resistance and source voltage. This circuit is also called current mirror or current limiter. This circuit possesses very high internal resistance than load resistance.

Figure 1.4 shows the transistorized constant current source circuit.

This constant current source provides current stabilization and large resistance which provides stable operating point and high emitter resistance to the differential amplifier. In Fig. 1.4 the base and collector of the transistor Q_1 are connected together to form a diode. The transistors Q_1 and Q_2 are same. Both collector and base are connected to the supply voltage $+V_{CC}$ through resistance R.

$+V_{CC}$ makes the transistor base–emitter junction forward biased and base–collector junction reversed biased as it is a *npn* transistor, thus it will work in active region.

From circuit, the current flowing through resistor R is I_R

$$\therefore \quad I_R = \frac{V_{CC} - V_{BE}}{R} \tag{1.24}$$

The forward cut in voltage $V_{BE} \approx 0.7\text{V}$

$$\text{Since} \quad I_R = I_C + 2I_B$$
$$= \frac{V_{CC} - V_{BE}}{R} \tag{1.25}$$

However $\beta = \frac{I_C}{I_B}$, *current gain* of the transistor. Substituting value of I_B in Eq. (1.25), we get I_C

Fig. 1.4 Constant current source

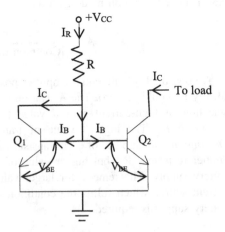

$$\therefore \quad I_C = \frac{V_{CC} - V_{BE}}{R} \left(\frac{\beta}{2 + \beta} \right) \tag{1.26}$$

If $\beta = 100$, then $\left(\frac{\beta}{2+\beta} \right)$ is very small, moreover change in the β value will not affect the value of bracket term to larger extent.

Hence the collector current I_C totally depends on V_{CC} and R values. As both transistors are identical, the I_C values in both transistors are same. The above equation indicates that $I_C \approx I_R$ and it is constant over a wide range of current gain value β of the transistor. The value of bracket term is nearly unity, and for a change in β value the collector current I_C, i.e. I_R will not change effectively. Hence collector current I_C will remain constant.

In case of transistor Q_2 is disconnected, the collecter current in transisor Q_1 equals to I_R and the I_R value, i.e. I_C value is set only by suppl voltage $+V_{CC}$ and R values. Hence the collector current in both transistors is same. Therefore it creates the image of collector current in transisor Q_2 as in transistor Q_1. Hence it is also called current mirror.

Constant current bias

Constant current bias provides current stabilization an stable operating point to the diferential amplifier. The desired value of DC emitter current I_E can be set up using constant current bias. The function is same as a constant current source. The circuit of constant current bias for differential amplifier is shown in Fig. 1.5.

In differential amplifier emitter resistance R_e is replaced by constant current bias. The DC collector current I_{C_3} is determined by resistors R_1, R_2 and R_E.

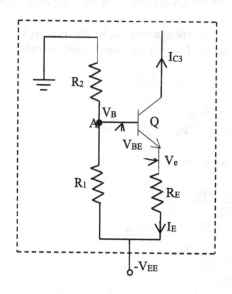

Emitter resistor R_e in $\approx$ differential amplifier is replaced by current source

Fig. 1.5 Constant current biasa

Applying voltage divider rule at point A, the base voltage V_B will be voltage across R_2.

$$\text{Hence} \quad V_B = \frac{R_2(-V_{EE})}{R_1 + R_2}$$

$$= -\frac{R_2 V_{EE}}{R_1 + R_2} \quad \text{and} \tag{1.27}$$

$$V_B = V_{BE} + V_e$$

$$\text{So} \quad V_e = V_B - V_{BE}$$

$$= -\frac{R_2 V_{EE}}{R_1 + R_2} - V_{BE} \tag{1.28}$$

$$\therefore \quad I_E = I_{C_3} = \frac{V_e - (-V_{EE})}{R_E}$$

$$I_{C_3} = \frac{V_{EE} + \left[-\frac{R_2 V_{EE}}{R_1 + R_2} - V_{BE}\right]}{R_E} \tag{1.29}$$

Thus the collector current I_{C_3} depends on the resistor values R_1, R_2, R_E and V_{EE} only, these values are fixed, hence collector current I_{C_3} is constant, and it provides constant current to the two halves of the differential amplifier. Also it provides very high resistance since current source has infinite internal resistance.

The *thermal stability* of transistor can be improved by connecting diodes D_1 and D_2 in place of resistor R_1. The diodes hold the emitter current constant even though there is a temperature change. Figure 1.6 shows the improved version of the circuit shown in Fig. 1.5.

To calculate the emitter current I_E, let the voltage drop across the diode is V_D and I_2 is the current flowing through resistor R_2. The transistor base voltage at point A is

$$V_B = -V_{EE} + 2V_D \quad \text{and} \tag{1.30}$$

$$V_e = V_B - V_{BE}$$

$$= -V_{EE} + 2V_D - V_{BE} \tag{1.31}$$

The current flowing through resistor R_E is

$$I_E = \frac{V_e - (-V_{EE})}{R_E}$$

$$= \frac{-V_{EE} + 2V_D - V_{BE} + V_{EE}}{R_E} \tag{1.32}$$

$$I_E = \frac{2V_D - V_{BE}}{R_E}$$

Fig. 1.6 Constant current
bias with diodes
compensation

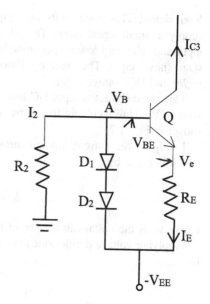

Assuming that diodes and transistors have same characteristics and $V_D \approx V_{BE}$, then emitter current I_E will be

$$I_E = \frac{V_D}{R_E} \qquad (1.33)$$

Thus the emitter current depends on the emitter resistor R_E and voltage drop across diodes. The change in temperature causes change in the emitter current I_E which will change the diode current but diode current is a part of current flowing through resistor R_2. Hence the current flowing through base of transistor will change. Ultimately the diode current compensates the base I_B to change emitter current I_E. Thus this will provide temperature compensation to bias the transistor thermally. For these two diodes if a suitable zener diode is connected in place of diodes, the circuit will work more efficiently by maintaining the zener current.

1.6   DC Level Shifter

In Op-Amp a number of differential stages are cascaded without coupling capacitors. Therefore the output DC level is always higher than the input DC level. This may build the output DC level towards positive supply voltage causes amplitude limitation and affects the linearity. So to maintain linearity, it is necessary to have quiscent operating state of the first stage differential amplifier. In order to reduce the output voltage closed to zero without input signal, the circuit employed is called *DC level shifter*. This stage shifts the output DC voltage level towards the negative supply in small signal input. The advantage is that the DC level shifter has high input and relatively low output impedance which prevents the loading of the gain stage (next stage). The *emitter follower* or *common collector* stage can serve as *buffer* and *DC voltage shifter*.

This stage shifts the input DC level V_i to the more negative DC level of output voltage V_0. Thus to obtain the output voltage V_0, consider the following circuit as shown in Fig. 1.7.

Let I_E be the current flowing through emitter of the transistor Q_1 (Fig. 1.7a) which is given by

$$I_E = \frac{V_E}{R_1 + R_2} \tag{1.34}$$

where V_E is the voltage at emitter of the transistor Q_1.

Applying voltage divider rule at output (Fig. 1.7a), hence output voltage V_0 will be

$$V_0 = \frac{R_2}{R_1 + R_2} V_E \tag{1.35}$$

Since $V_E = V_i - V_{BE}$

$$\therefore \quad V_0 = \frac{(V_i - V_{BE})R_2}{R_1 + R_2} \tag{1.36}$$

The voltage drop across R_1 can be increased for obtaining the desired DC voltage shift. This simple circuit has main disadvantage that AC gain starts

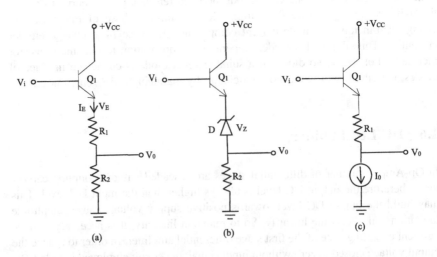

Fig. 1.7 a–c Level shifter circuits

decreasing as R_2 decreases, because $\frac{R_2}{R_1+R_2}$ provides the voltage attenuation. Thus to remove this difficulty, resistor R_1 in Fig. 1.7a can be replaced by *zener diode or avalanche diode* (Fig. 1.7b). So the net DC level shift is given by

$$V_0 - V_i = -(V_{BE} + V_Z) \tag{1.37}$$

where V_Z is the breakdown voltage of Avalanche diode D.

Another way is to replace R_2 by constant current source I_0 as shown in Fig. 1.7c. The DC level shift is given by

$$V_0 - V_i = -(V_{BE} - I_0 R_1) \tag{1.38}$$

Because of constant current source there is no attenuation in voltage signal since it has very high resistance.

An alternative DCdc level shifter with V_{BE} muliplier utilized in IC Op-Amps is shown in Fig. 1.8.

The voltage V_{AB} between the collector and emitter of Q_2 is given by

$$V_{AB} = V_{BE}\left(1 + \frac{R_1}{R_2}\right) \tag{1.39}$$

So it is called V_{BE} multiplier.

However output voltage V_0 will be

$$V_0 = V_i - V_{BE} - V_{AB} \tag{1.40}$$

Hence

$$V_0 - V_i = -V_{BE}\left[2 + \frac{R_1}{R_2}\right] \tag{1.41}$$

In such stage small signal voltage gain A_V is unity, so it is simply a voltage follower. The DC level shift can be acurately adjusted by proper selection of R_1 and R_2 ratio.

1.7 Output Stage

The output stage of Op-Amp is designed in such a way to provide the substantial power to the load. Also it should have low output resistance, large output current and voltage swing capability as good as total supply voltage ($V_{CC} + V_{EE}$) peak to peak. All these features posesses in *complementry emitter follower* which may be class-B or class-AB configuration. Figure 1.9 shows the class-AB output stage complementry emitter follower.

Fig. 1.8 DC level shifter
with V_{BE} multiplier

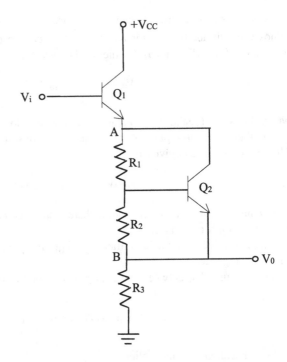

During posivite half of the first sinusoidal cycle transistor Q_1 conducts and acts as a source of supply of current to the load resistor R_L, during this time transistor Q_2 becomes off. When negtive half cycle comes, the pnp transistor Q_2 conducts and supplies the current to the load resistor R_L, at this time transistor Q_1 is off.

But this circuit has failure due to crossover distortion. However this can be eliminated by introducing two p–n junction diodes as connected in the circuit (Fig. 1.10).

The bias voltage applied is greater than the cut in voltage V_γ of the diodes, i.e. $V > 2V_\gamma$. In this circuit by employing the diodes D_1 and D_2 crossover distortion is eliminated, but the transfer characteristic not passes through origin hence when $V_i = 0$, V_0 is finite ($V_0 \neq 0$).

Thus alternative way is to employ the V_{BE} multiplier in place of diodes D_1 and; so that it eliminates the crossover distortion as well as satisfies the condition when $V_i = 0$, then V_0 is almost zero. The circuit with V_{BE} multiplier is shown in Fig. 1.11. In this case both transistors Q_1 and Q_2 are conducted slightly under quiescent condition.

The overload protection is provided in the circuit of output stage. Sometimes if the ambient temperature of the transistors increases by the excessive power dissipation or the short circuit of the output to the common or power supply terminal, the possibility of damaging the transistors increases. To avoid this damage, the current limiting resistors are connected internally in series with both the transistors. Various types of current limiting circuits are available as resistive, diode clamp output, transistor clamp output current limiting circuits, etc.

Fig. 1.9 Complementary
emitter follower

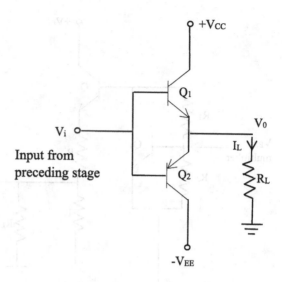

Fig. 1.10 Class B output
stage

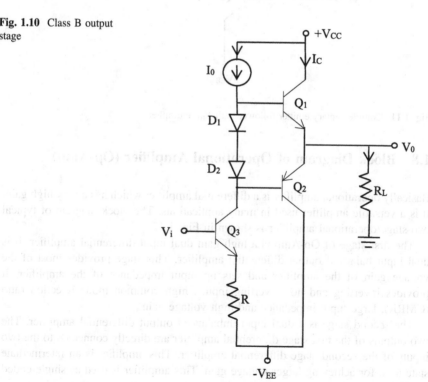

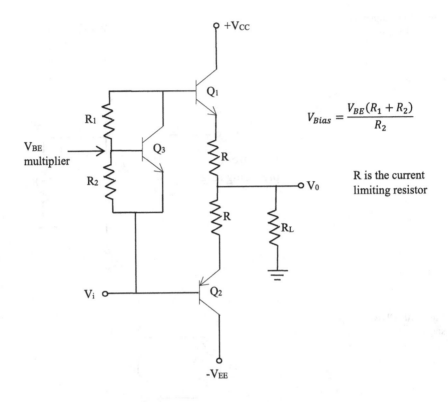

$$V_{Bias} = \frac{V_{BE}(R_1 + R_2)}{R_2}$$

R is the current limiting resistor

Fig. 1.11 Complementary emitter follower with V_{BE} multiplier

1.8 Block Diagram of Operational Amplifier (Op-Amp)

Basically operational amplifier is a differential amplifier which has a very high gain. It is a versatile amplifier used in many applications. The block diagram of typical two-stage operational amplifier is shown in Fig. 1.12.

The first stage of Op-Amp is a high gain dual input differential amplifier. It is dual input balanced output differential amplifier. This stage provides most of the voltage gain of the amplifier and sets up input impedance of the amplifier. It provides inverting and non-inverting inputs, high common mode rejection ratio (CMRR), large input impedance and high voltage gain.

The second stage is a dual input unbalanced output differential amplifier. The two outputs of the first stage differential amplifier are directly connected to the two inputs of the second stage differential amplifier. This amplifier is an intermediate state used for achieving larger voltage gain. This amplifier is used as single-ended amplifier.

Third state is a level shifter that adjusts the DC voltages or shifts the DC voltages down to zero with respect to ground. This is called *level shifter or level translator*.

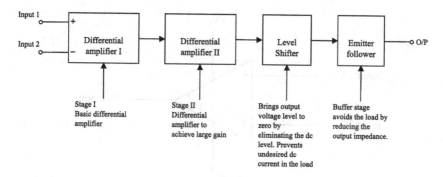

Fig. 1.12 Block diagram of operational amplifier

The first and second stages are direct coupled, thus the DC voltage levels at the output of intermediate or second stage differential amplifier are somewhat above the ground. So to bring that level to zero volts in absence of any input signal, the level shifter is used.

The final stage is the emitter follower. It is an amplifier having high input impedance and very low output impedance, as well as voltage gain is unity. This stage acts as a buffer between the preceding and the next stage. It avoids the unavoidable loading.

Thus Op-Amp has two input terminals and one output terminal. One of the input terminal is called inverting input, and other is called non-inverting input.

Signal appearing at the output inverts or is out of phase with input which is designated as *inverting input* denoted by −ve sign.

Signal appearing at the output is in phase with the input which is designated as *non-inverting input* denoted by +ve sign.

The output voltage is directly proportional to the difference of the input voltages applied to two input terminals, i.e. $V_0 \propto (V_1 \sim V_2)$ or $A = \frac{V_0}{(V_1 \sim V_2)}$, where V_1 and V_2 are the inputs and V_0 is the output. The constant of proportionality A is called voltage gain of the amplifier.

The symbol of Op-Amp is shown in Fig. 1.13.

Input 1 is inverting, and input 2 is non-inverting input terminal. This amplifier requires dual power supply as $\pm V_{CC}$.

This amplifier gives nonzero output in the absence of input signal. To make this voltage to zero before application, two offset terminals are provided by which the output voltage can be adjusted to the ground potential. In good-quality and high precision Op-Amp ICs, this provision is already made in a level shifter circuit, but in some ICs, e.g. 741, we have to make such provision externally by connecting a 10 kΩ potentiometer between two offset null terminals, and variable terminal of potentiometer is connected to $-V_{CC}$. By adjusting the resistance of the poten-tiometer with input terminals connected to ground, the output voltage would be adjusted to zero volts.

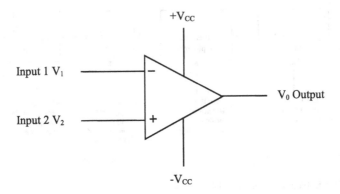

Fig. 1.13 Symbol of operational amplifier

The monolithic integrated circuits available in market are of different manu-facturers. Fairchid, Motorola, National semiconductor, Signetics, RCA, Texas instruments, Burr Brown, Intersil, Siliconix, etc., are the leading manufacturers of integrated circuits. Generally they manufacture all types of ICs including general-purpose and digital ICs. Op-Amp IC is a linear IC having analog circuit. Op-Amp ICs are classified into two groups as general purpose and special purpose. General-purpose ICs are cheap, and special-purpose ICs are costlier. IC 741 is a general-purpose IC used for variety of applications; whereas ICs like LM 380 or LM 381 are the special-purpose ICs used for specialized applications. The details regarding characteristics and applications are given in the linear IC data manual.

The most general-purpose IC is 741 having features such as short-circuit pro-tection, no frequency compensation, offset voltage null capability, large common mode and differential voltage ranges, low power consumption and no latch-up. IC 741 is available in three package types as 10-pin flat pack, 8-pin metal can and 8 or 14 pin DIP. This IC works in temperature range 0 to +70 °C. The power supply requirement is ±15 V.

1.9 The Single-Stage Architecture Op-Amp

To achieve large gain and large bandwidth, two-stage differential amplifiers are used, but these features can be made available in single-stage architecture Op-Amps. The single-stage Op-Amp has less gain and large bandwidth than multistage amplifier, and multistage amplifier has more gain and less bandwidth so that the gain-bandwidth product remains constant. Hence single-stage Op-Amps are more advantageous than the multistage amplifier. Figure 1.14 shows the block diagram of single-stage architecture Op-Amp. These configurations are commer-cially used.

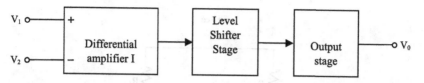

Fig. 1.14 Block diagram of single-stage architecture Op-Amp

High-speed Op-Amps such as the Op-Amps having unity gain-bandwidth say 15 MHz and slew rate of 50 V/µs are single-stage architecture. This type of Op-Amp consists of differential amplifier (single-stage only), level shifter and output stage. The first stage of differential amplifier employed the *Darlington pair* in place of single transistor. The *cascode* configuration of transistors gives larger bandwidth at a given gain other than common emitter amplifiers called *CE-CB* amplifier. This configuration offers large bandwidth, high gain, high slew rate, large input impedance and more stability. It improves input–output isolation and eliminates *Miller effect* which ultimately improves the bandwidth of the amplifier. The Darlington pair differential amplifier is shown in Fig. 1.15.

The current gain β of the *composite transistor* or Darlington pair can be calculated as follows. A single composite transistor structure is shown in Fig. 1.16, where two transistors Q_1 and Q_2 are used as a single transistor.

I_{b1} and I_{b2} are the base currents of transistors Q_1 and Q_2, respectively. Similarly I_C is the collector current which divides into I_{C1} and I_{C2} called output current. Hence

$$I_C = I_{C1} + I_{C2} \tag{1.42}$$

The emitter current flowing through transistor Q_1 is the input or base current of transistor Q_2.

$$\text{Thus } I_{C2} = \beta\, I_{b2} \quad \left(\because \frac{I_C}{I_b} = \beta\right) \tag{1.43}$$

where β is the current gain of the transistor.

$$\therefore \quad I_{C2} = \beta(1+\beta)I_{b1} \quad [\because I_{b_2} = (1+\beta)I_{b1}] \tag{1.44}$$

Substituting values of collector currents I_{C1} and I_{C2} in Eq. (1.42), we get

$$I_C = \beta I_{b1} + \beta(1+\beta)I_{b1} \quad \text{Since } I_{C1} = \beta I_{b1}$$
$$= \beta I_{b1} + \beta I_{b1} + \beta^2 I_{b1}$$
$$= (2\beta + \beta^2)I_{b1} = \beta(2+\beta)I_{b1} \tag{1.45}$$

$$\text{or} \quad \frac{I_C}{I_{b_1}} = \beta^2 \quad (\because 2+\beta \approx \beta \text{ since } \beta \gg 2)$$

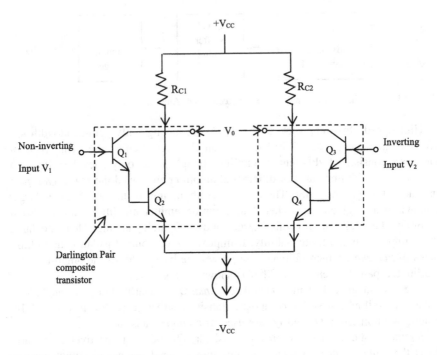

Fig. 1.15 Darlington pair differential amplifier

Fig. 1.16 Darlington pair transistor

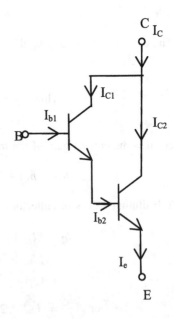

In most of the transistors the value of β is greater than 100. Therefore the current gain of the composite transistor or Darlington pair is collector current divided by input current.

$$\beta_0 = \frac{I_C}{I_{b_1}} = \beta^2 \tag{1.46}$$

If the value of β is considered as 100 as in most of the transistors, the current gain of Darlington pair would be 10,000. This indicates that the current gain is enhanced enormously.

If Darlington pair is used as emitter follower, then its gain becomes unity but input resistance increases and output resistance decreases extremely. As we have discussed in AC analysis (Eq. 1.19) the input resistance

$$R_{i2} = R_{i1} = 2\beta_{ac}re \tag{1.47}$$

Merely using Darlington pair the $\beta \rightarrow \beta^2$ which will increase the input resistance of the differential amplifier than the single transistor. Also the common mode gain A_c reduced to a very small value as compared to the single transistor in differential amplifier.

However if CE-CB *(cascode)* configuration is used instead of composite transistor, the high-frequency response of the amplifier get improved, which increases the bandwidth of the amplifier.

1.10 Electrical Characteristics of an Ideal Operational Amplifier (Op-Amp)

An ideal Op-Amp has following *electrical characteristics.*

1. Infinite gain $A = \infty$
2. Infinite input impedance $Z_i = \infty$
3. Zero output impedance $Z_0 = 0$
4. Zero output voltage for $V_d = 0$, i.e. zero offset
5. Infinite bandwidth BW $= \infty$
6. Common mode rejection ratio is infinite, i.e. CMRR $= \infty$
7. Infinite slew rate
8. Input bias current zero
9. Input offset current zero.

The typical practical values of electric parameters are as follows:

1. Open loop gain $(A) \geq 10^4$
2. Input impedance $Z_i \approx$ High $\geq 10^6 \, \Omega$

3. CMRR $\geq Z_0$ dB
4. Output impedance Z_0 low $< 500\ \Omega$
5. Input bias current—low
6. Input offset voltage—low < 10 mV
7. Input offset current—low < 0.2 nA.

1.11 Equivalent Circuit of an Ideal Op-Amp

The equivalent circuit of an *ideal Op-Amp* is shown in Fig. 1.17.

In the equivalent circuit the symbols have their usual meaning. V_1 and V_2 are the input voltages at inverting and non-inverting terminals; A is the open loop gain, Z_i is the input impedance, and Z_0 is the output impedance of an ideal Op-Amp. $\pm V_{CC}$ are the supply voltages, and V_0 is the output voltage.

The input voltage V_d is the difference voltage between two input terminals of an Op-Amp, thus

$$V_d = A(V_1 - V_2) \tag{1.48}$$

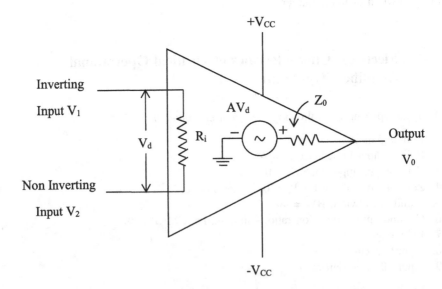

Fig. 1.17 Equivalent circuit of an ideal Op-Amp

The output voltage V_0 is directly proportional to difference between the two input voltages.

$$\text{Hence} \quad V_0 = A(V_1 - V_2) = AV_d \tag{1.49}$$

Chapter 2
Feedback in Amplifiers

The concept of *feedback* in amplifiers is equally important. Without feedback design, amplifier is worthless. So as to get better understanding of both types of feedbacks, positive and negative, the theory of feedback in amplifier is discussed in this chapter. The negative and positive feedbacks have their own advantages. In a negative feedback amplifier the bandwidth, stability, input impedance increase, whereas the application of positive feedback produces oscillations. Not only have these systems had the advantages but few disadvantages also. The details are discussed in respective chapters. The operational amplifier has infinite gain (ideal); hence it could not be operated in open loop condition, otherwise it would go into saturation caused damage to the IC. Therefore we have to always operate the Op-Amp in closed loop condition. This closed loop condition is nothing but the feedback amplifier. Introduction, types of negative feedback, non-inverting amplifier, i.e. voltage series feedback and effect of feedback on various important properties, viz. closed loop voltage gain, input impedance, output impedance, frequency response, etc., is reported. The voltage follower circuit is also discussed. Similarly the concept of voltage shunt feedback in amplifier, i.e. inverting amplifier and its effect on the above-discussed properties is reported in this chapter. The concept of virtual ground and sign changer is also discussed.

2.1 Introduction

The sampling of the output of the amplifier fed back to the input via feedback network is called feedback. The feedback network may be a passive network. The feedback signal combines with the external signal source by means of mixer. This combined signal is the input of the amplifier.

In feedback amplifiers loop gain is important. For the calculation of the loop gain various methods have been employed by the researchers (Sedra et al. and

© The Author(s), under exclusive license to Springer Nature Singapore Pte Ltd. 2022
S. Yawale and S. Yawale, *Operational Amplifier*,
https://doi.org/10.1007/978-981-16-4185-5_2

Rosenstark). Mostly two methods have been suggested by many authors (Hurst and Hurst). The most useful method adopted is two-port analysis and return ratio method, but the loop gain obtained by these two methods is entirely different and does not agree with each other. The reason being given by authors for the disagreement is forward signal transfers through feedback network which is generally ignored at low frequencies, but it cannot be neglected at high frequencies. The performance of the amplifier depends on the frequency response, and frequency response is totally depending on the feedback provided to the amplifier.

A simplified analysis of feedback amplifiers is nicely discussed by Jose Luis in his paper. However in this book a very general and simple concept of feedback is used to understand the readers.

There are two types of feedback.

(a) Negative feedback and b) positive feedback.

2.1.1 Negative Feedback

If the signal fed back to the input is out of phase by 180° with input signal, the feedback is called *negative feedback*. This feedback reduces the output voltage or gain of the amplifier; therefore it is called *degenerative feedback*.

The effect of negative feedback is to reduce the gain, increase the band width (keeping gain-bandwidth product constant with and without feedback), reduce distortion and increase the stability. Also the input impedance of the amplifier increases, and output impedance reduces. These are the advantages of negative feedback in amplifier.

2.1.2 Positive Feedback

In *positive feedback* the output signal fed back to the input is in phase with input signal (0° or 360°). The feedback signal adds to the input signal; hence it is called *regenerative feedback*. This type of feedback is used in oscillators. In such type of feedback overall voltage gain of the amplifier increases, and also it increases the instability and distortion.

Consider the closed loop system with feedback as shown in Fig. 2.1. Let A be the open loop gain of the amplifier and β is the feedback gain.

Let V_i be the input voltage and V_f be the feedback fraction, then the difference in the voltage

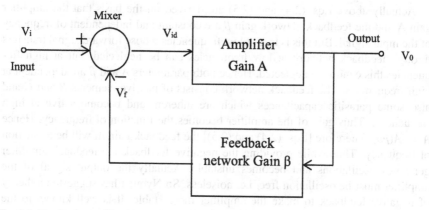

Fig. 2.1 Feedback amplifier block diagram

$$V_{id} = V_i - V_f \tag{2.1}$$

and

$$V_f = \beta V_0 \tag{2.2}$$

where β is the feedback factor.

The output voltage V_0 is

$$V_0 = AV_{id} \tag{2.3}$$

$$V_0 = A(V_i - V_f)$$
$$= A(V_i - \beta V_0)$$
$$\text{or } (1 + A\beta)V_0 = AV_i$$

$$\therefore \frac{V_0}{V_i} = \frac{A}{(1+A\beta)} \left(\because A_f = \frac{V_0}{V_i} \right) \text{ where } A_f \text{ is gain with feedback.}$$

$$A_f = \frac{A}{(1+A\beta)} \quad \text{for negative feedback} \tag{2.4}$$

and

$$A_f = \frac{A}{(1-A\beta)} \quad \text{for positive feedback} \tag{2.5}$$

Actually above Eqs. (2.4) and (2.5) are derived on the basis that the amplifier gain A and the feedback network gain β are constant and independent of frequency of the input signal. But this is true at low frequencies whose forward signal transfers through feedback network at low frequencies can be neglected, but at high frequencies this cannot be neglected. Hence both parameters A and β are dependent at high frequencies. The feedback network consists of passive elements R and C and also some parasitic capacitances which are inherent and become active at high frequencies. Thus gain of the amplifier becomes the function of frequency. Hence $A \rightarrow A(j\omega)$. Therefore [Eqs. (2.4) and (2.5)] the feedback gain A_f will be a function of frequency. Thus while providing the negative feedback the feedback amplifier generates oscillations and becomes unstable. Actually the output signal of the amplifier must be oscillation free, i.e. noiseless. So Nyquist has suggested a theory of negative feedback to make the amplifier more stable. It is well known to the researchers working in feedback systems of amplifiers the *Nyquist criteria*.

So Eqs. (2.4) and (2.5) can be written as

$$A_f(j\omega) = \frac{A(j\omega)}{(1 + A(j\omega)\beta)} \tag{2.6}$$

In Eq. (2.1), if $1 + A(j\omega)\beta = 0$, then $A_f(j\omega) \rightarrow \infty$ means $A(j\omega)$ has finite value. That is without any input, there will be a finite output signal. So the amplifier becomes unstable. Thus oscillations will be generated at the output of some frequency ω_o. It means that the Eq. $1 + A(j\omega_o)\beta = 0$ is satisfied. *Nyquist criteria* tells that polar plot of $A(j\omega_o)\beta$ has -1 value.

$$\text{i.e. } 1 + A(j\omega)\beta = 0 \quad (\because A(j\omega)\beta = -1) \tag{2.7}$$

As *feedback network* gain β is constant, then

$$\begin{aligned} |A(j\omega)\beta| &= 1 \text{ and} \\ \text{arc tan } A(j\omega) &= -180° \end{aligned} \tag{2.8}$$

Negative sign in phase angle indicates the lags of the output signal with input signal.

2.2 Negative Feedback in Operational Amplifier

There are four types of configurations of negative feedback. These four types are derived from the series and shunt or parallel connection of feedback network with input and output of amplifier.

(a) Voltage series negative feedback
(b) Voltage shunt negative feedback
(c) Current series negative feedback
(d) Current shunt negative feedback.

The most commonly used feedbacks such as voltage series and voltage shunt are important. Hence these are discussed in this book. Readers might be interested in current series and current shunt negative feedback should refer the other reference book.

2.3 Non-inverting Amplifier (Voltage Series Feedback Amplifier)

Operation of Op-Amp in non-inverting mode is simply a voltage series feedback, rather voltage series feedback is employed in non-inverting Op-Amp. The circuit diagram of non-inverting amplifier is shown in Fig. 2.2. The circuit output is taped through resistance R_f called feedback resistor and resistor R_1. Input signal is given through inverting input of Op-Amp. Voltages V_1 and V_2 are the voltages at inverting and non-inverting terminals of Op-Amp. Supply voltage $\pm Vcc$ are already connected from dual power supply, to analyse this amplifier by evaluating various parameters such as closed loop voltage gain, input and output impedance, output offset voltage and unity gain-bandwidth.

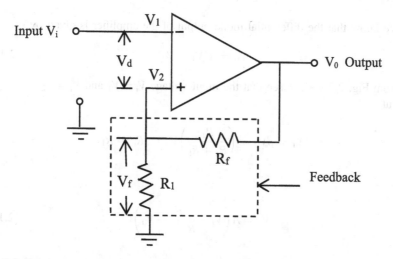

Fig. 2.2 Non-inverting Op-Amp or voltage series feedback amplifier

2.3.1 Closed Loop Voltage Gain

Let V_i be the input from signal source having internal resistance zero given to non-inverting terminal of Op-Amp. The feedback network consists of feedback resistor R_f and resistor R_1. The voltage across resistor R_1 is

$$V_f = \left(\frac{R_1}{R_1 + R_f}\right) V_0 \Rightarrow V_f = \beta V_0 \quad \text{or } \beta = \frac{V_f}{V_0} \left(\because \beta = \frac{R_1}{R_1 + R_f}\right) \qquad (2.9)$$

The open loop gain of the amplifier is denoted by

$$A = \frac{V_0}{V_d} \qquad (2.10)$$

The closed loop gain of the amplifier is denoted by

$$A_f = \frac{V_0}{V_i} \qquad (2.11)$$

Applying Kirchhoff's voltage law for the input loop (Fig. 2.2).
Thus

$$V_i = V_d + V_f$$

or

$$V_d = V_i - V_f \qquad (2.12)$$

where V_f is the feedback voltage and V_d is the difference input voltage, i.e. ($V_1 - V_2$).

We know that the differential mode output of the amplifier is given by

$$V_0 = A(V_1 - V_2) \qquad (2.13)$$

From Fig. 2.2 we can see that the input voltage $V_i = V_1$ and $V_f = V_2$.
But

$$V_f = \beta V_0 = \left(\frac{R_1}{R_1 + R_f}\right) V_0 \quad \because R_i \gg R_1$$

Hence

$$V_0 = A\left(V_i - \frac{R_1}{R_1 + R_f} V_0\right) \qquad (2.14)$$

On rearranging the above Eq. (2.14), we get

$$\frac{v_0}{V_i} = \frac{A(R_1 + R_f)}{R_1 + R_f + AR_1} \qquad (2.15)$$

Since open loop gain A is very large (Ideal case $A = \infty$)

$$\therefore \quad AR_1 \gg (R_1 + R_f) \text{ and } R_1 + R_f + AR_1 \approx AR_1$$

Thus

$$\frac{v_0}{V_i} = A_f = 1 + \frac{R_f}{R_1} \qquad (2.16)$$

Equation (2.16) gives the gain of the amplifier with feedback
or

$$A_f = \frac{R_1 + R_f}{R_1}$$
$$= \frac{1}{\beta} \qquad (2.17)$$

On rearranging Eq. (2.15), we get
Divide denominator and numerator by $(R_1 + R_f)$, the closed loop gain A_f will be

$$A_f = \frac{\left(\frac{R_1 + R_f}{R_1 + R_f}\right)A}{\frac{R_1 + R_f}{R_1 + R_f} + \frac{AR_1}{R_1 + R_f}}$$

$$A_f = \frac{A}{1 + A\beta} \qquad (2.18)$$

Equation (2.18) gives the relationship between open loop gain A, closed loop gain A_f and feedback factor β.

Equation (2.16) suggests that the closed loop gain of the non-inverting amplifier depends on feedback resistor R_f and resistor R_1 only. It is independent of open loop gain. Also it seems that the gain is greater than unity and independent of temperature changes and supply voltage changes. The closed loop gain is independent of frequency means the gain of the amplifier is same for DC as well as high frequencies.

From Eq. (2.10), we can write the open loop gain as

$$A = \frac{V_0}{V_d}$$

or

$$V_d = \frac{V_0}{A}$$

But $A = \infty$ in ideal case, hence $V_d \approx 0$ and the differential input $V_d = V_1 - V_2$.
Hence

$$V_1 = V_2 \tag{2.19}$$

This shows that the voltages at inverting and non-inverting terminals are same means differential input voltage is ideally zero.

2.3.2 Input Impedance

The equivalent circuit for the voltage series feedback of Op-Amp is shown in Fig. 2.3.

The resistance seen from the input terminals when negative feedback is applied in non-inverting Op-Amp is called *input impedance* in closed loop circuit, denoted as Z_{if}

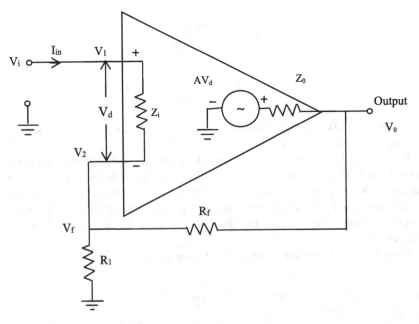

Fig. 2.3 Equivalent circuit for voltage series feedback

$$Z_{if} = \frac{V_i}{I_{in}}$$

$$\text{But} \quad I_{in} = \frac{V_d}{Z_i} \tag{2.20}$$

$$\text{Thus} \quad Z_{if} = \frac{V_i}{V_d/Z_i}$$

From Eq. (2.10), $V_d = \frac{V_0}{A}$ and

$$V_0 = \left(\frac{A}{1+A\beta}\right) V_i$$

$$Z_{if} = Z_i \frac{V_i}{V_0/A} \tag{2.21}$$

Substituting value of V_o in above Eq. (2.21)

$$\therefore \quad Z_{if} = Z_i \frac{AV_i}{\left(\frac{A}{1+A\beta}\right) V_i} \tag{2.22}$$

$$= Z_i(1+A\beta)$$

This shows that the input impedance with feedback Z_{if} is $(1 + A\beta)$ times original input impedance Z_i of the amplifier. The open loop gain A of the Op-Amp is very large ideally $A = \infty$; hence input impedance with feedback will be more than the input impedance without feedback Z_i.

2.3.3 Output Impedance

The equivalent circuit for the calculation of output impedance of Op-Amp in non-inverting closed loop circuit is shown in Fig. 2.4. While deriving the expression for output impedance, the non-inverting input is directly connected to the ground and a source is applied externally at the output terminal and the ground to calculate the resulting current I_o. Once output current is known the output impedance can be calculated. Therefore we determine the output current I_o.

Thus the output impedance Z_{of} with feedback is

$$Z_{of} = \frac{V_o}{I_o} \tag{2.23}$$

The *output impedance* is the impedance seen from the output terminals M and ground.

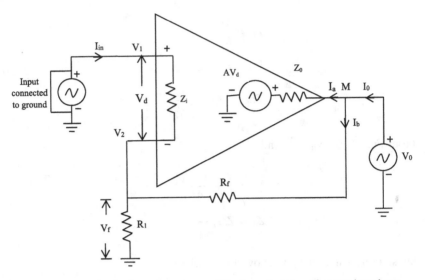

Fig. 2.4 Equivalent circuit of non-inverting amplifier for calculation of output impedance

Applying Kirchhoff's current law at node M, we get, resulting current I_o as

$$I_o = I_a + I_b \tag{2.24}$$

From equivalent circuit we can observe that resistance R_1 parallel to input impedance Z_i, i.e. $R_1 \parallel Z_i$ and feedback resistance R_f is in series, i.e. $(R_f + R_1 \parallel Z_i) > Z_0$.

Hence $I_a > I_b \therefore I_0 = I_a$.

Applying Kirchhoff's voltage law for the output loop

$$V_0 - Z_0 I_0 - AV_d = 0$$
$$\text{So} \quad I_0 = \frac{V_0 - AV_d}{Z_0} \tag{2.25}$$

$$\text{But} \quad V_d = V_1 - V_2 = -V_f$$
$$= -\frac{R_1}{R_1 + R_f} V_0$$
$$= -\beta V_0 \quad \therefore \beta = \frac{R_1}{R_1 + R_f}$$

Substituting value of V_d in Eq. (2.25), we get

$$I_0 = \frac{V_0 + A\beta V_0}{Z_0}$$

$$= \frac{(1 + \beta A)V_0}{Z_0} \qquad (2.26)$$

Substituting the value of I_0 in Eq. (2.23), we get

$$Z_{0f} = \frac{V_0}{\frac{(1 + \beta A)V_0}{Z_0}}$$

$$Z_{0f} = \frac{Z_0}{(1 + A\beta)} \qquad (2.27)$$

Equation (2.27) shows that the output impedance of non-inverting amplifier is 1/ $(1 + A\beta)$ times the original output impedance of the amplifier. Since A is very large, the output impedance of the amplifier becomes much smaller than the Z_0. In other words the output impedance of the amplifier reduces by the negative feedback.

2.3.4 Frequency Response

It is known that the gain of the amplifier varies with input signal frequency. Generally in RC-coupled amplifiers, the voltage gain falls at lower and higher frequencies and remains constant over mid-frequency range or band of frequencies. The frequency range or a band of frequencies over the gain remains constant and is called *bandwidth* of the amplifier.

The product of the constant gain and bandwidth is called gain-bandwidth product. This *gain-bandwidth product* in any amplifier with and without feedback remains constant means if gain increases, bandwidth decreases and if bandwidth increases, gain of the amplifier decreases, so that the product before and after feedback remains constant.

Generally the variation of gain with frequency is plotted, from which bandwidth and cut-off frequencies can be calculated and is called frequency response. For most general-purpose Op-Amp IC 741, the break frequency f_o is 5 Hz. It is the frequency at which gain falls or rolls of 20 dB/decade. The frequency response of the IC 741 is shown in Fig. 2.5.

If break frequency is 5 Hz and gain is 2×10^5, then gain-bandwidth product is 1 MHz or if gain is 100 the frequency f_o is 10 kHz, then product is 1 MHz. Hence the *break frequency* at which the gain obtained is unity is called *unity gain-bandwidth*.

In the Op-Amp having single break frequency, the unity gain-bandwidth product can be written as

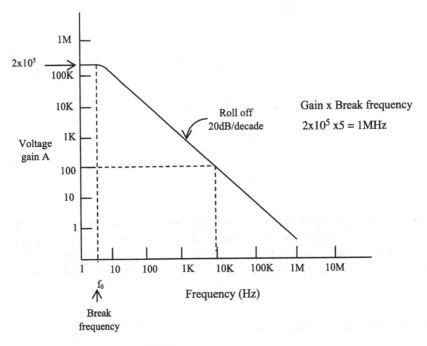

Fig. 2.5 Frequency Response of IC 741

$$\text{Unity gain bandwidth} = Af_0 \tag{2.28}$$

$$\text{or with feedback the unity gain bandwidth} = A_f f_f \tag{2.29}$$

Equating Eqs. (2.28) and (2.29), we get

$$Af_0 = A_f f_f$$

or

$$f_f = \frac{Af_0}{A_f} \tag{2.30}$$

However in case of non-inverting amplifier with feedback the closed loop gain

$$A_f = \frac{A}{1 + A\beta}$$

Substituting value of A_f in Eq. (2.30), we get

$$f_f = f_0(1 + A\beta) \tag{2.31}$$

where f_0 is the break frequency of Op-Amp, A is the open loop gain, β is the feedback circuit gain, and f_f is the bandwidth with feedback.

Above Eq. (2.31) shows that the bandwidth with feedback in non-inverting amplifier is break frequency f_0 times $(1 + A\beta)$. As A is the open loop gain of Op-Amp which is very large, the bandwidth increases with negative feedback.

2.3.5 Total Output Offset Voltage (V_{OT})

In ideal case the output of the Op-Amp should be expected to be zero when no input signal is applied. However because of mismatching of transistors, some small output voltage is present. This output voltage is called *output offset voltage* (V_{OT}). This output offset voltage will change due to negative feedback in amplifiers. The maximum value of this voltage can swing up to $\pm V_{CC}$ in open loop system. Thus the total output offset voltage (V_{OT}) is given by (with negative feedback).

$$V_{OT} = \frac{\pm V_{CC}}{(1 + A\beta)} \tag{2.32}$$

This indicates that the total output offset voltage with negative feedback in non-inverting amplifier reduces by a factor of $1/(1 + A\beta)$.

Hence negative feedback in non-inverting Op-Amp reduces gain, increases bandwidth, increases input impedance and reduces output impedance of the amplifier. Not only these parameters change but the noise will also get reduced. All these parameters enhance the quality of the amplifier. Hence negative feedback in Op-Amp is important.

2.4 Voltage Follower or Unity Gain Buffer

The voltage follower or unity gain buffer or isolation amplifier has unity gain means there is no amplification of signal, and the output is exactly same as input so it is called *voltage follower*. It is the amplifier like emitter follower in transistor amplifier. This unity gain buffer possesses very high input and very low output impedance.

The non-inverting amplifier can be converted into voltage follower by applying total output voltage as feedback. Therefore feedback resistance $R_f = 0$ and $R_1 = \infty$, i.e. open. The circuit will look like as shown in Fig. 2.6.

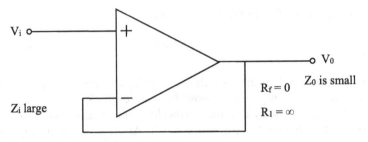

Fig. 2.6 Voltage follower

Now we can calculate the various parameters such as voltage gain, input impedance, out impedance, bandwidth and total output offset voltage.

2.4.1 Voltage Gain (A_f)

The voltage gain A_f of the amplifier in non-inverting mode with feedback is

$$A_f = 1 + \frac{R_f}{R_1} \tag{2.33}$$

Substituting $R_f = 0$ and $R_1 = \infty$ in Eq. 2.33, we get

$$A_f = 1 \text{ (Unity)} \tag{2.34}$$

Thus the gain of the amplifier with feedback or closed loop gain in non-inverting mode is unity.

2.4.2 Input Impedance (Z_{if})

The input impedance Z_{if} is calculated in previous Sect. (3.2) which is given by

$$Z_{if} = Z_i(1 + A\beta) \quad \because \beta = \frac{R_1}{R_1 + R_f} \tag{2.35}$$

Substituting $R_f = 0$ and $R_1 = \infty$, the feedback circuit gain, $\beta = 1$. Hence Eq. (2.35) becomes

$$Z_{if} = AZ_i \tag{2.36}$$

Equation (2.36) indicates that the input impedance of a voltage follower increases by open loop gain times the original input impedance Z_i. In other words the input impedance becomes very large.

2.4.3 Output Impedance (Z_{of})

The output impedance Z_{of} in non-inverting amplifier is given by (Sect. 2.3.3)

$$Z_{of} = \frac{Z_0}{(1+A\beta)} \qquad (2.37)$$

Since $\beta = 1$, $Z_{of} = \frac{Z_0}{(1+A)}$.

or simply

$$Z_{of} = \frac{Z_0}{A} \quad (\because 1+A \approx A) \qquad (2.38)$$

It means that the output impedance Z_{of} in voltage follower reduces by a factor of $1/A$.

2.4.4 Bandwidth or Frequency Response (f_f)

The bandwidth with feedback is given by (Sect. 3.4)

$$f_f = f_0(1+A\beta) \qquad (2.39)$$

Since $\beta = 1, f_f = f_0(1+A) = f_0 A \quad (\because 1+A \approx A) \qquad (2.40)$

Equation (2.40) shows that the bandwidth of a voltage follower increases by a factor A.

2.4.5 Total Output Offset Voltage (V_{OT})

This voltage is given by (Sect. 2.3.5)

$$V_{OT} = \frac{\pm V_{sat}}{(1+A\beta)} \qquad (2.41)$$

$$\text{Since } \beta = 1, V_{OT} = \frac{\pm V_{\text{sat}}}{(1+A)} = \frac{\pm V_{\text{sat}}}{A} \qquad (2.42)$$

Equation (2.42) shows that the total output offset voltage (V_{OT}) reduces by a factor $1/A$. Since open loop gain $A = \infty$, the output offset voltage V_{OT} reduces drastically. This amplifier is called buffer because it is connected between two stages to avoid the loading or is used in impedance matching circuit as shown in Fig. 2.7.

If we have the voltage divider network as in Fig. 2.7, then to avoid the loading effect this voltage follower or buffer can be used. Without buffer this voltage divider has the impedance $R_1\|R_2$ which directly becomes parallel with load resistance R_L, i.e. $R_1\|R_2\|R_L$. This effective resistance will change the voltage appearing across resistance R_1. That is it will load the divider circuit. The remedy is to connect voltage follower between resistance R_1 of voltage divider and load resistance. This will allow a high impedance source to drive low impedance load keeping voltage across resistor R_1 as it is due to the unity gain of buffer. In a very general concept the voltage follower can be introduced as buffer between two networks.

2.5 Inverting Amplifier (Voltage Shunt Feedback Amplifier)

This is one of the simplest and commonly used configurations. Figure 2.8 shows the circuit diagram of inverting amplifier with feedback. A voltage shunt feedback is used in this circuit.

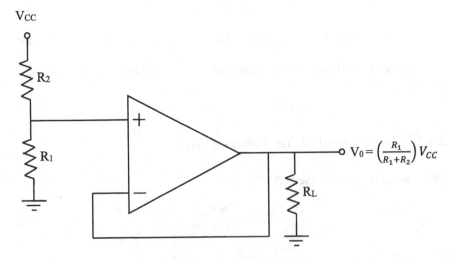

Fig. 2.7 Simple circuit of voltage divider for avoiding loading effect

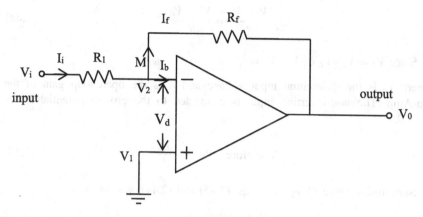

Fig. 2.8 Inverting amplifier with voltage shunt feedback

In inverting amplifier feedback resistor is connected between inverting input and output. The input signal is provided through resistance R_1 to the inverting terminal of the Op-Amp. The non-inverting input is grounded. Now let us evaluate various parameters such as closed loop voltage gain, virtual ground, input impedance, output impedance, frequency response and output offset voltage.

2.5.1 Closed Loop Voltage Gain

Applying Kirchhoff's current law to the input loop of the amplifier

$$\text{So} \quad I_i = I_b + I_f \tag{2.43}$$

where I_i is the input current that flows through the resistor R_1 and I_f is the current flows through the feedback resistor R_f. I_b is the current entering into inverting input of the Op-Amp.

But input impedance of the ideal Op-Amp is infinite; hence current I_b does not enter into the Op-Amp, i.e. very negligible current flows through the inverting input of Op-Amp.

Hence

$$I_i = I_f \quad (\because I_b \approx 0)$$

Let us calculate I_i, the current flowing through R_1 and I_f, current flowing through feedback resistance R_f. Using ohm's law in Fig. 2.8.

$$\frac{V_i - V_2}{R_1} = \frac{V_2 - V_o}{R_f} \tag{2.44}$$

Since $V_d = V_1 - V_2$ or $V_1 - V_2 = \frac{V_o}{A}$

where V_d is the differential input voltage and A is the open loop gain of the Op-Amp. The non-inverting input is connected to the ground potential hence $V_1 = 0$.

$$\text{Therefore} \quad V_2 = -\frac{V_o}{A} \tag{2.45}$$

Substituting value of V_2 from Eqs. (2.45) and (2.44), we get

$$\frac{V_i - \left(-\frac{V_o}{A}\right)}{R_1} = \frac{-\frac{V_o}{A} - V_o}{R_f}$$

On rearranging the terms to obtain the ratio $\frac{V_o}{V_i}$ as

$$A_f = \frac{V_o}{V_i} = -\frac{AR_f}{R_1 + R_f + AR_1} \tag{2.46}$$

where A_f is the gain of amplifier with feedback. The open loop gain A of the Op-Amp is very large, hence

$$AR_1 \gg (R_1 + R_f)$$

$$\therefore \quad A_f = -\frac{AR_f}{AR_1}$$

$$A_f = -\frac{R_f}{R_1} \tag{2.47}$$

Equation (2.47) shows that voltage gain of inverting amplifier depends on the feedback resistor R_f and input resistor R_1 values.

The negative sign indicates that output voltage is 180° out of phase with input signal voltage.

By the proper selection of R_f and R_1, the gain of the amplifier can be adjusted to desired value. The closed loop gain of the amplifier is constant and independent of amplitude and frequency; hence no amplitude distortion occurs and the frequency response is wide because R_f and R_1 are linear components. Thus this amplifier will operate as excellent linear device.

The closed loop gain can be obtained in terms of open loop gain. Divide Eq. (2.46), by $(R_1 + R_f)$ denominator and numerator to

$$A_f = -\frac{\frac{AR_f}{(R_1 + R_f)}}{1 + \frac{AR_1}{(R_1 + R_f)}}$$

$$= -\frac{A\mu}{1 + A\beta} \tag{2.48}$$

$\because \beta$ is the feedback circuit gain equal to $\frac{R_1}{(R_1 + R_f)}$ and μ is called attenuation factor equal to $\frac{R_f}{(R_1 + R_f)}$. On comparison of closed loop gain in non-inverting amplifier (Eq. 2.18) with Eq. (2.48), it is observed that the closed loop gain of inverting amplifiers is attenuation factor μ the closed loop gain of non-inverting amplifier.

2.5.2 Virtual Ground

Virtual ground means it is not physically connected to ground potential but virtually we can consider inverting terminal in Op-Amp as connected to ground potential. The voltage at point M in Fig. 2.8 is virtually at ground potential. This would be clear from the following discussion.

(i) *To calculate the potential difference between node M and ground*
   We know that open loop gain of the amplifier

$$A = -\frac{V_0}{V_2}$$

   or

$$V_2 = -\frac{V_0}{A} \tag{2.49}$$

   Since open loop gain $A = \infty$ (ideal case), thus the potential difference V_2 between node M and ground is zero.

(ii) *To calculate impedance between node and ground*

   The current entering at point M is I_i and potential difference is V_2, thus the impedance at node M will be

$$Z(M) = \frac{V_2}{I_i} = \frac{V_2}{I_f} \quad (\because I_i = I_f)$$

Substituting the value of I_f from Eq. (2.43 to 2.49) we get,

$$Z(M) = \frac{V_2}{\frac{(V_2 - V_0)}{R_f}}$$

$$= \frac{R_f}{1 - \frac{V_0}{V_2}}$$

$$Z(M) = \frac{R_f}{1 - A} \left(\because \frac{V_0}{V_d} = A, V_d = V_1 - V_2 \text{ or } V_d = V_2 \right) \text{ Since } V_1 = 0 \quad (2.50)$$

As open loop gain $A = \infty$ (ideal case)

$Z(M) = 0$

Thus the effective impedance $Z(M)$ at point M is zero. It means that V_1 and V_2 terminals are virtually shorted. Therefore this point M is called virtual ground.

2.5.3   Input Impedance (Z_{if})

Since the input impedance between node M and ground is zero, hence

$$Z_i = R_1 + Z(M)$$
$$= R_1 + 0$$

$$\therefore Z_i = R_1 \quad (2.51)$$

Hence effective impedance of the amplifier in inverting mode is totally decided by input resistance R_1 externally connected in the circuit.

The input impedance of the amplifier with feedback is the effective resistance seen from input side of the amplifier. We cannot select any value of resistance R_1 because if we select smaller value, the effective resistance becomes small which will come across the output will load the amplifier. This will draw more current from the Op-Amp and will damage the IC.

2.5.4   Output Impedance (Z_{of})

The output impedance of the amplifier with feedback is the resistance seen from the output side. The same expression as in non-inverting mode is there because instead of inverting terminal being directly grounded it is grounded by shorting the source in equivalent circuit and non-inverting terminal is directly grounded.

The output current I_0 can be measured by connecting source externally between output terminal and ground. Hence treatment is same as in non-inverting mode amplifier. Thus from previous equation (Refer non-inverting amplifier Sect. 2.3)

$$Z_{of} = \frac{Z_0}{1+A\beta} \qquad (2.52)$$

where Z_0 is the original output impedance of Op-Amp, A is the open loop gain and β is the feedback network gain. Equation (2.52) indicates that the output impedance with feedback of the amplifier is $\frac{1}{1+A\beta}$ times original impedance Z_0. Thus output impedance reduces by a factor of $\frac{1}{1+A\beta}$. Since ideal gain $A = \infty$, the output impedance Z_{of} reduces exceedingly.

2.5.5 Frequency Response

The frequency response is same as discussed in non-inverting mode. The bandwidth of the amplifier with feedback is given by

$$f_f = f_0(1+A\beta) \qquad (2.53)$$

where symbols have usual meaning as in non-inverting mode frequency response.

2.5.6 Total Output Offset Voltage (V_{OT})

The output offset voltage with feedback is given by (Refer non-inverting Sect. 2.3)

$$V_{OT} = \frac{\pm V_{CC}}{1+A\beta} \qquad (\because \pm V_{sat} = \pm V_{CC}) \qquad (2.54)$$

Equation (2.54) indicates that output offset voltage reduces with feedback in inverting mode amplifier.

2.6 Sign Changer or Inverter

Following Fig. 2.9 shows the circuit of *sign changer or inverter*. In inverting amplifier if feedback resistor R_f and input resistance R_1 are equal, then gain of the amplifier will be

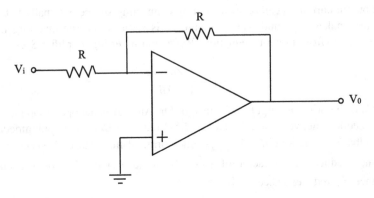

Fig. 2.9 Sign changer or inverter

$$A_f = -\frac{V_0}{V_i}$$

$$= -\frac{R_f}{R_1}$$

(2.55)

Since $R_f = R_1$, $A_f = -1$ or $V_0 = -V_i$.

This means that the output of the amplifier is equal in magnitude of input signal but with opposite sign.

This is the special case of inverting amplifier.

Chapter 3
Parameters of Operational Amplifier and Instrumentation Amplifier

The parameters of Op-Amp are significant to make it most successful integrated circuit. The *parameters* such as input bias current, input offset current, input offset voltage, open-loop gain, common mode rejection ratio (CMRR), supply voltage rejection ratio (SVRR), output resistance, slew rate and frequency response are some of the foremost parameters to make this device versatile. CMRR, SVRR or power supply rejection ratios are important parameters while designing the circuit of operational amplifier. Slew rate is also equally important to select the Op-Amp IC for specific application. The fast response of the circuit depends on the change in the output voltage with change in input signal voltage. For example sample and hold circuit, analog-to-digital converter (ADC) and digital-to-analog converter (DAC), teleswitching, data acquisition system, digital communication and signal analysers, etc., require extremely sensitive operational amplifiers having fast response. Other parameters too have the significance in the fabrication or designing of high gain circuits. In view of this, proper selection of Op-Amp IC from the manufacturer's data sheet can be made. Mostly low-cost ICs are preferred for general purpose where parameters have less significant value. The ultrasensitive ICs are costlier and used for specific applications.

3.1 Introduction

The parameters of Op-Amp are significant to make it most successful integrated circuit. The *parameters* such as input bias current, input offset current, input offset voltage, open-loop gain, common mode rejection ratio (CMRR), supply voltage rejection ratio (SVRR), output resistance, slew rate and frequency response are some of the foremost parameters to make this device versatile. CMRR, SVRR or power supply rejection ratios are important parameters while designing the circuit of operational amplifier. Slew rate is also equally important to select the Op-Amp IC for specific application. The fast response of the circuit depends on the change in the output voltage with change in input signal voltage. For example sample and hold circuit, analog-to-digital converter (ADC) and digital-to-analog converter (DAC), teleswitching, data acquisition system, digital communication and signal analysers, etc., require extremely sensitive operational amplifiers having fast response. Other parameters too have the significance in the fabrication or designing of high gain circuits. In view of this, proper selection of Op-Amp IC from the manufacturer's data sheet can be made. Mostly low-cost ICs are preferred for general purpose where parameters have less significant value. The ultrasensitive ICs are costlier and used for specific applications.

These parameters are extremely useful while designing instrumentation amplifiers also. Instrumentation amplifier should be very sensitive to the weak signals. The reproduction of exact amplified signal is the essential property of any amplifier. Less noise, less distortion and more stability in the amplifier contribute to the quality of the instruments. Larger gain and more bandwidth of the amplifier are significant properties. All these good properties come through the various parameters of the Op-Amp. For example the ICs used in biomedical instrumentation amplifier require extremely high input impedance and very high input impedance creates the noise at the input level. To suppress this noise (may be generated by AC mains, wireless devices, lights, mains, etc.), you have to select an amplifier having specific properties. This can be selected from the values of the various parameters of the Op-Amp. Mostly medical instruments are used in hospital where signal generated from biopotential electrodes will be of the order of microvolts and noise level may be in the range of millivolts. In that case we have to select such a device having better noise suppression, less input bias current, high CMRR etc. Therefore, the AD82X series of instrumentation amplifier would be selected for the design of biomedical equipment such as ECG and EEG. Not only it is applicable in biomedical instrumentation but also in industrial applications we require special features of Op-Amp. Especially it is needed for the weak signal amplification. So design engineer has to go first for the data manual of the Op-Amp ICs for proper selection. The special-purpose ICs are more costly than the general purpose because additional features incorporated in the IC increase the cost.

3.2 Parameters of Op-Amp

3.2.1 Input Bias Current

In ideal case input as well as the output of an Op-Amp should be zero. The input of individual transistors in the Op-Amp circuit gets properly biased but there will be some base current flowing through the input transistors even in the absence of any input signal, i.e. when input is equal to zero. The input current at the two inputs is the base current in the first differential amplifier stage of Op-Amp which is one half of the sum of separate current flowing into the input terminals when output is equal to zero (see Fig. 3.1).

$$\text{Hence } I_B = \frac{I_{B1} + I_{B2}}{2} \quad \text{when } V_0 = 0 \tag{3.1}$$

Smaller the input bias current, larger would be the input impedance of the Op-Amp. Generally, FET can reduce input bias current because of its high input impedance.

3.2.2 Input Offset Current

Actually perfectly matched transistors are used in differential amplifier so base currents would be exactly equal.

The difference between the separate input current flowing into the input transistors when output is zero volts is called *input offset current*.

$$\text{Hence, } I(i0) = I(B_1) \sim I(B_2) \text{ when } V_0 = 0$$

Smaller the input offset current better would be the Op-Amp.

3.2.3 Input Offset Voltage

In ideal case the output $V_0 = 0$ when there is no input. But in practical use the output is nonzero even if input is zero, i.e. $V_1 = 0 = V_2$. This may be due to mismatch of the base to emitter voltage in input differential amplifier transistors. This nonzero output can be made zero by applying input of proper polarity.

The input voltage that must be applied between the two input terminals to get output zero is called *input offset voltage* (see Fig. 3.2).

Lesser the input offset voltage, better would be the Op-Amp.

Fig. 3.1 Input bias current

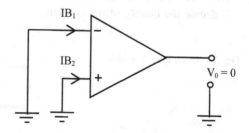

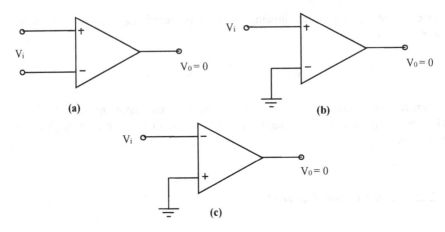

Fig. 3.2 Input offset voltage

3.2.4 *Open Loop Gain*

It is the ratio of output voltage to the voltage applied between two inputs of Op-Amp (Fig. 3.3).

$$A = \frac{V_0}{V_i} \tag{3.2}$$

3.2.5 *Common Mode Rejection Ratio (CMRR)*

Common mode rejection ratio (CMRR) is the ratio of differential mode gain A_d to the common mode gain A_c.

$$\text{CMRR} = \left|\frac{\text{Differential mode gain}}{\text{Common mode gain}}\right| = \left|\frac{A_d}{A_c}\right| \tag{3.3}$$

Generally, common mode gain AC should be small, therefore CMRR would become large. Higher the value of CMRR better would be the amplifier. CMRR will decide the quality of the Op-Amp.

Fig. 3.3 Open-loop amplifier

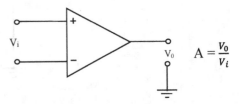

The common mode and differential mode gain configuration is shown in Fig. 3.4a, b, respectively.

3.2.6 Differential Input Resistance (R_i)

Differential input resistance (R_i) is the equivalent resistance that can be measured at either the inverting or non-inverting inputs with the other terminal connected to ground. The typical value of R_i is 10^{12} Ω for μA741 IC.

3.2.7 Input Capacitance (C_i)

Input capacitance (C_i) is the equivalent capacitance that can be measured at either the inverting or non-inverting input terminal with the other terminal connected to ground. The typical value of C_i is 1.4 pf for μA741 IC.

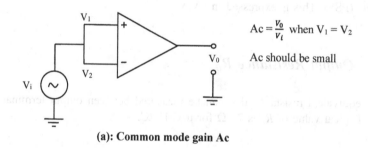

$Ac = \dfrac{v_0}{v_t}$ when $V_1 = V_2$

Ac should be small

(a): Common mode gain Ac

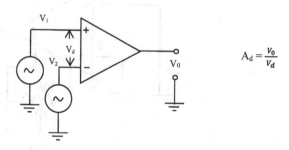

$A_d = \dfrac{v_0}{v_d}$

(b): Differential mode gain A d

Fig. 3.4 **a** Common mode gain A_c, **b** differential mode gain A_d

3.2.8   Offset Voltage Adjustment

When $V_1 = V_2 = 0$, i.e. both inputs are equal to zero, we get same nonzero output. To adjust this output equal to zero, the offset null terminals are provided to the Op-Amp IC. In case of IC μA741 pin numbers 1 and 5 are offset null terminals. The 10 KΩ potentiometer is connected between these pins 1 and 5 and variable terminal to the negative supply $-V_{EE}$. By varying the potentiometer the output offset voltage can be reduced to zero volts (see Fig. 3.5).

3.2.9   Supply Voltage Rejection Ratio (SVRR)

Supply voltage rejection ratio (SVRR) is defined as the change in Op-Amp input offset voltage V_{offset} due to change in supply voltage V.

$$\therefore \text{SVRR} = \frac{\Delta V_{\text{offset}}}{\Delta V} \qquad (3.4)$$

This is also called *power supply rejection ratio* (PSRR) or the *power supply sensitivity* (PSS). This is expressed in μV/V.

3.2.10   Output Resistance R_0

It is the equivalent resistance that can be measured between output terminal and ground. Typical value of R_o is 75 Ω for μA741 IC.

Fig. 3.5 Offset voltage adjustment

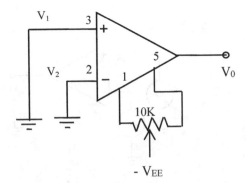

3.2.11 Slew Rate (SR)

It is defined as the rate of change of output voltage per unit time which is expressed in volts per μs.

$$\therefore SR = \frac{d(V_0)}{dt}\bigg|_{max} \qquad (3.5)$$

Slew rate indicates the response of Op-Amp to the change in input frequency. Slew rate limits the performance of the operational amplifier. It is an important parameter. If the slew rate is exceeded, the output of the operational amplifier can be distorted and there will be no faithful output of the amplifier. In many applications where speed is the requirement, the slew rate can have a significant effect.

3.2.12 Frequency Response

The gain of the Op-Amp is almost constant for all frequencies which considered up till now. But the gain is a function of frequency. Thus for a specific frequency the gain will have specific magnitude as well as a phase angle. Hence variation in frequency will cause variation in gain magnitude and its phase angle. The Op-Amp which responds to different frequencies and changes its gain is called *frequency response*.

A plot of gain versus frequency is called *frequency response plot* (Fig. 3.6). The gain is generally expressed in decibels (dB) and frequency in Hz units.

The gain A(dB) versus frequency and phase angle versus frequency plots are called *Bode plot*. In *Bode plot*, gain is always represented in dB which is generally used for determination of stability and network design.

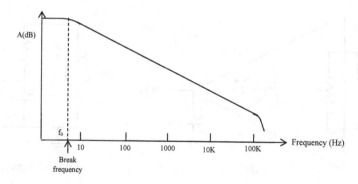

Fig. 3.6 Frequency response

Break frequency (f_o)

It is the break point frequency (f_o) at which the voltage gain starts to falls off or the point at which frequency starts roll-off (−3 dB point).

3.3 Open-Loop Voltage Gain as a Function of Frequency

Following Fig. 3.7 represents high-frequency model of the Op-Amp with a single break frequency. The high-frequency model is a modified version of the equivalent circuit of Op-Amp. Simply capacitor C is added at the output.

Let us obtain an expression for the gain as a function of frequency. So applying voltage divider rule, we get,

$$V_0 = \frac{-jX_c}{R_0 - jX_c} AV_d \qquad (3.6)$$

Since $-j = 1/j$ and capacitive reactance $X_c = \frac{1}{2\pi fC}$

$$\therefore V_0 = \frac{1/j2\pi fC}{R_0 + 1/j2\pi fC} AV_d$$
$$= \frac{AV_d}{1 + j2\pi fR_0C} \qquad (3.7)$$

Hence open-loop voltage gain is

$$A(f) = \frac{V_0}{V_d} \qquad (3.8)$$

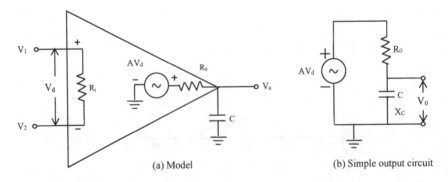

(a) Model   (b) Simple output circuit

Fig. 3.7 a, b High-frequency model of Op-Amp with single break frequency

$$= \frac{A}{1 + j2\pi f R_0 C}$$

$$= \frac{A}{1 + \left(f/f_0\right)} \quad \left(\because f_0 = \frac{1}{2\pi R_0 C}\right) \tag{3.9}$$

where $A(f)$ is open-loop voltage gain as a function of frequency
A is the gain of the Op-Amp at DC
f is the operating frequency and
f_0 is the break frequency of Op-Amp.

Equation (3.9) represents open-loop gain which is complex quantity so this gain can be expressed in polar form as

$$|A(f)| = \frac{A}{\sqrt{1 + \left(f/f_0\right)^2}} \tag{3.10}$$

and phase angle

$$\phi(f) = -\tan^{-1}\left(\frac{f}{f_0}\right) \tag{3.11}$$

Both gain and phase angle are functions of frequency. Using Eq. (3.10) we can obtain gain versus frequency plot and from Eq. (3.11) phase angle versus frequency plot, which are shown in Fig. 3.8.

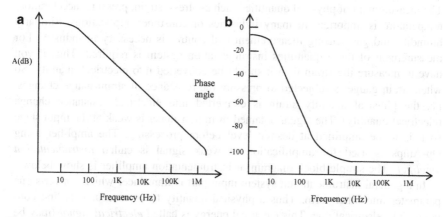

Fig. 3.8 a Gain versus frequency plot and b phase angle versus frequency plot

Equation (3.10) can be represented in dB as

$$20 \log |A(f)| = 20 \log \left[\frac{A}{\sqrt{1 + \left(f/f_0 \right)^2}} \right] \tag{3.12}$$

$$A(f)dB = 20 \log A - 20 \log \sqrt{1 + \left(f/f_0 \right)^2} \tag{3.13}$$

From above equation we can conclude that—

(1) The open-loop gain $A(f)$ dB is approximately constant from 0 Hz to the break frequency f_0
(2) When input signal frequency $f = f_0$ the break frequency, the gain $A(f)$ dB is 3 dB down from its value at 0 Hz so the break frequency is sometimes called the −3 dB *frequency or corner frequency.*
(3) The open-loop gain $A(f)$ dB is approximately constant up to *break frequency* f_0, but it decreases 20 dB for increase of one decade in frequency. So gain rolls off at a rate of 20 dB/decade.
(4) The specific value of input signal frequency at which the $A(f)$ dB is zero is called *unity gain bandwidth product (UGB).*

3.4 Instrumentation System

The measurement of physical quantities such as stress, strain, pressure, acceleration, temperature is important in many industries or consumer applications. Similarly humidity and gas sensing measurement and control is necessary in industry. For measurement of these quantities instrumentation system is required. Thus if you have to measure the strain then it should be converted into electrical quantity. So when strain gauge is subjected to pressure the resistance of strain gauge changes, i.e. the physical quantity strain is converted into electrical resistance change (electrical quantity). The signal obtained from transducer is weak at the input stage so it is to be amplified at desired level before processing. The amplifier using Op-Amps is used for amplification of weak signal is called *instrumentation amplifier.* The simple block diagram of instrumentation amplifier is show below.

In a general instrumentation system input part is transducer which converts one parameter into other form. Thus a physical quantity to be measured is first converted into electrical one. This electrical energy is called *electrical signal* (may be weak) further processed under instrumentation amplifier to become strong having sufficient strength to drive next circuit. The weak signal obtained from transducer is not sufficient to drive the further electronic circuit due to weak strength. Hence we always use the amplifier to strengthen the transducer output signal. The device which does this function is called instrumentation amplifier. This instrumentation amplifier has some specific properties such as high input impedance, high gain and low output impedance, more stability and noise-free signal amplification. All these properties are of course associated with operational instrumentation amplifier. Therefore instrumentation amplifier is emphasized to elaborate instrumentation system. In any instrumentation system the first part is transducer. So there are various kinds of transducers. Some of them are listed below-

a. Mechanical
b. Electrical
c. Chemical
d. Thermal
e. Acoustic
f. Hydraulic
g. Optical
h. Biomedical.

For example microphone is an acoustic transducer which converts acoustic energy, i.e. sound waves into electrical energy. Another example is speaker which performs reverse function of *microphone*. The electrical energy is converted into sound energy. These are the examples of acoustic transducer. Similarly we can take the example of *chemical transducer* as *gas sensor*. A gas sensor consists of some metal oxides or composites or conducting polymers that can be used to detect the concentration of ambient gas. Such type of *gas sensor* converts gas concentration to the change in electrical resistance of the sensor. This change in electrical resistance of the sensor is used to change the electrical signal by any means. Then we would be able to process this signal change due to change in gas concentration using instrumentation amplifier. The *transducers* available are of both types reversible and irreversible. Battery is the well-known example of reversible-type transducer that converts chemical energy into electrical one and vice versa whereas optical transducers are irreversible.

Actually the transducer is to be selected has some accuracy, sensitivity, stability, repeatability and ruggedness. These are the main characteristics that transducers should have. Generally the instrumentation system can be classified into three ways-

a. Analog instrumentation system
b. Digital instrumentation system
c. Biomedical instrumentation system.

Let us discuss these instrumentation systems in brief-

3.4.1 Analog Instrumentation System

The simple block diagram of analog instrumentation system is shown in Fig. 3.9. This system consists of three blocks named as-

i. Input stage
ii. Intermediate stage and
iii. Output stage.

The input stage consists of input transducer where *physical quantity* to be measured or processed for further work is connected. This stage has transducer and preamplifier. The function of this stage is the physical quantity to be measured that is converted into its proportional electrical quantity or signal. This electrical signal may be changed in resistance, voltage, current, capacitance, inductance or any other means of electrical signal required to be processed.

The intermediate stage consists of instrumentation amplifier for increasing the signal strength coming from previous stage, processor as modulation/ demodulation unit frequency selection, filter, wave shaping, transmission and conversion, etc. The intermediate stage is very important as well because of nature of the signal and strength or power of the signal which drives the next stage. As per the processing of the signal and its nature, the devices such as filters, analog-to-digital converters and modulators are used.

Generally the signal which can be processed in a proper way having accuracy, stability and reliable (distortion less) signal is fed to the next stage. The third and final stage is an indicator and in some instrumentation system automatic process controller. These are generally indicating units such as cathode ray oscilloscope (CRO), chart strip recorder, X–Y recorder, printer, electronic memory devices, computer data acquisition system, lamps or indicating instruments like voltmeter, digital multimeter and speakers can be used but it requires driving circuits. Now mostly digital systems are used therefore data will be stored on the hard disc of the computer for further use and processing.

While developing any system points to remember that the instrumentation system should have

(a) Good accuracy
(b) Better resolution
(c) Fidelity and
(d) Large signal-to-noise ratio.

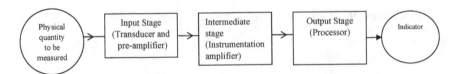

Fig. 3.9 Block diagram of general instrumentation system

3.4.2 Digital Instrumentation System

The basic structure of instrumentation system is same but requirements are different. As this is a *digital instrumentation system,* initially we have to convert the analog signal to digital one which machine can understand. Conversions, processing and controlling of the signal and then obtained that signal in readable form are the useful and important steps in the digital instrumentation system. The block diagram of digital instrumentation system is shown in Fig. 3.10.

(a) *Sensor/transducer*

The input available may be in analog form having any value and this analog input or signal will be varying continuously. Therefore before being sent to the processing, it is necessary to convert it into such a form which the machine can understand. Hence it is very important to convert any type of analog signal (of course physical quantity) into electrical signal by means of transducer or sensor or combination of both. The sensor has prime importance because it forms the input part of the system. Thus proper selection of the sensor having selective parameters or characteristics such as greater sensitivity is necessary. Hence while designing the instrumentation the design engineer has to choose the proper sensor for that particular system.

(b) *Signal conditioning*

Signal conditioning circuits are the important one because the output of the sensor or transducer cannot be directly connected to the circuits like analog-to-digital converter (ADC). If it becomes possible then it will be an ideal instrumentation system where the output of ADC can be available in readable form. Generally the electrical output of the sensor/transducer is weak and has noise too. This obtained signal is not sufficient to drive the next circuit. In such cases we have to amplify the signal that requires preamplifier, remove the noise that requires filter circuit prior being data send to next circuit such circuits are called *signal conditioners.* These are called *signal manipulators* also. Here inverting and non-inverting mode operational amplifier can be utilized.

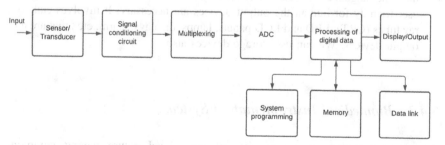

Fig. 3.10 Digital instrumentation system

(c) *Multiplexing*

Many times data is available in multiple signals. This data we have to handle at a time and send it to the next circuit for conversion. So the multiplexing technique is the best option for handling time-varying data. *Multiplexing* is a process to handle the combined multiple signals/data into one signal. The device used for multiplexing data is called *multiplexer*. If the device handles analog signal is called *analog multiplexer* and if it handles digital data is called *digital multiplexer*. There are various types of multiplexing techniques such as time division multiplexing (TDM), frequency division multiplexing (FDM) and wavelength division multiplexing (WDM).

(d) *Analog-to-digital converter (ADC)*

Signal coming from analog multiplexer needs to be converted to the digital format. This work can be better way done by the *analog-to-digital converter* (ADC). The specific characteristics such as *sensitivity, resolution, accuracy* and *response time* are the important one while selecting the ADC. Good IC of ADC having special properties is nowadays available in the market. The data manual can help you the proper selection.

(e) *Data processing*

The output obtained from ADC is in binary format called *digital data*. This data is a raw data which can be processed for further rather it needs to be. So this data is sent to the *microprocessor or microcomputer* requires storing of data hence memory, processing of data by the programmer, then controlling of data, checking of data before and after processing, etc., such types of different steps are involved in processing unit. Moreover this data to be sent to the keyboard, display unit or output devices or to provide the data link, etc., is the common routines to the microprocessor. Also this should be handled by the software so end-to-end software linking and to generate proper correct output is the difficult and exhaustive task. It requires skill and intelligence of the designer as well as software engineer. A skilled designer or the physicist having sound knowledge of software as well as hardware can do this job.

(f) *Display/output*

The processed correct signal coming from the preceeding stage is in the form of machine language. This cannot be read by the human. To make it readable and user-friendly for human the output devices are necessary. It involves cathode ray tubes (CRT), LED and LCD panels, lamps, displays, alarm, etc., that are the output devices. We can use storage devices also.

3.4.3 Biomedical Instrumentation System

Any instrumentation system consists of generalized nature having common components as input device, intermediate stage (consists of signal conditioning, processing, etc.) and output stage. In bioinstrumentation system, extra care is taken to make the system more precise and accurate. Thus extra components are needed to make the system perfect.

In *human body* each cell acts like a small battery; therefore it generates the small potential of the order of millivolt (mV) or current of the order of microampere. This potential is called the biopotential used as a source for investigation of various diseases. For example electrocardiography (ECG), electromyography (EMG) and electroencephalogram (EEG) are totally based on the biosignal or biopotential. These signals are very sensitive and weak so common mode interference is always there to disrupt the signal. The ECG signal is generated due to electrical activity of heart whose amplitude is 10 microvolt to 5 mV and frequency range is 0.05–120 Hz. This type of weak signal is detected and processed in the instrumentation system in such a way that exact information is collected from the output. The block diagram of biomedical instrumentation system is shown in Fig. 3.11.

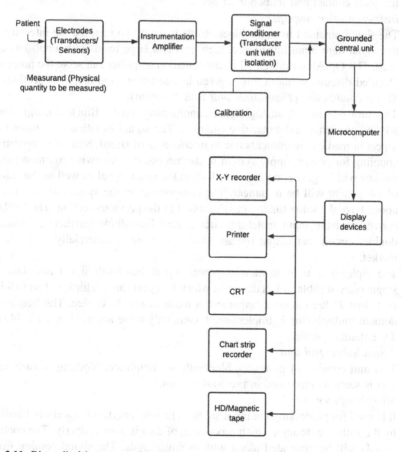

Fig. 3.11 Biomedical instrumentation system

The function of each block is discussed in brief. Readers who are interested in detailed description of the block should refer books of bioinstrumentation system.

(a) *Electrodes*

Electrodes are transducers that produce electric potential when placed on the various points of the human body. These are classified as (i) surface electrodes and (ii) deep-seated electrodes.

The electrodes that produce signal from the surface of the tissue are called surface electrodes.

When electrode is placed inside the live tissue or cell generates signal or potential is called deep-seated electrode.

Generally aluminium, copper, platinum, gold or silver are used as metal electrodes due to their good potential with respect to hydrogen electrode.

While fixing the electrode on the skin, the skin impedance is important. To reduce this impedance, electrolyte or jelly is used between electrode and skin for good contact and reduction of noise.

(b) *Instrumentation amplifier*

The signal obtained from the bioelectrode is weak. To increase the intensity of the signal high input impedance, high gain and large bandwidth amplifiers are used. The Op-Amp differential instrumentation amplifier can serve the purpose. Detailed discussion about this is given in instrumentation amplifier section.

(c) *Signal conditioner (Transducer unit with isolation)*

This unit consists of analog signal conditioning, ADC, Bluetooth, diplexers, RF/DC converter and capacitive isolators. The signal isolation is a major key aspect in medical instruments due to interference of signal. Similarly capacitive coupling for power supply system is also necessary. Otherwise common mode voltage will be generated and it will affect the main signal as well as the safety of the patient will be in danger. The components of the system as mentioned above have the same function as discussed in the previous section. The RF/DC converters have input matching circuit, zero bias diode rectifiers in voltage doubler circuit and output bypass filter. These are commercially available in market.

The diplexer is used to transfer power signal, bluetooth signal and data on single coaxial cable. It is a device in which two ports are multiplexed on to third port. Just a filter circuit arrangement is made inside the device. The frequency domain multiplexing is implemented. Generally these are used in TV cable or TV antenna systems.

(d) *Grounded central unit*

This unit consists of diplexers, bluetooth and amplifier. Working of each section is same as discussed in previous section.

(e) *Microcomputer*

It is used for processing the input signal. The relevant data is carefully handled by the software through which processing of data is done properly. The control signals will be generated along with timing signals. The signal coming from this section is fed to the display devices.

(f) *Calibration*

Calibration of the instruments is necessary without that the reliability of signal cannot be decided. Hence proper calibration is done in this section.

(g) *Display devices or output devices*

X–Y recorder, printer, CRT, chat strip recorder, hard disk, memory or magnetic tape or CD can be used as display or output devices. The working is same as in the previous section. It produces output which is understandable to the user.

The instrumentation amplifier is a part and parcel of the any instrumentation system. Therefore more importance and emphasis is given on Op-Amp instrumentation. None other than Op-Amp instrumentation amplifier is adoptive to the instrumentation system.

3.5 DC and AC Amplifiers

Op-Amp can amplify two types of signals, i.e. direct current (DC) and alternating current (AC). Both amplifiers are important as their application point view. In most of the cases only DC signal is present and AC signal has to suppress (noise), for example output of thermocouple or thermopile. On the other hand, in many applications AC signal is important and DC signal has to suppress, for example output of AC excited bridge. Therefore these amplifiers are discussed separately.

3.5.1 DC Amplifier

In DC amplifier the output signal changes in response to change in DC input signal. A DC amplifier can be inverting, non-inverting and differential. In DC amplifier, the offset null is necessary. Following Fig. 3.12 shows (a) inverting, (b) non-inverting and (c) differential DC amplifiers.

3.5.2 AC Amplifier

The inverting and non-inverting Op-Amp circuit responds to AC signals as well as DC signals. But if we have to limit the low and high frequency limits then use of coupling capacitors in AC amplifier is necessary. In audio receivers or public address system number of stages are connected or cascaded. Also due to coupling capacitors the DC signal prevented from amplification. Following Fig. 3.13 shows (a) AC-inverting and (b) non-inverting amplifiers with coupling capacitors. The coupling capacitor not only blocks the DC voltage but also sets the low frequency cut-off limit, which is given by (Fig. 3.14).

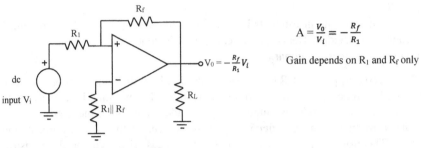

$$A = \frac{V_0}{V_i} = -\frac{R_f}{R_1}$$

Gain depends on R_1 and R_f only

(a): Inverting dc amplifier

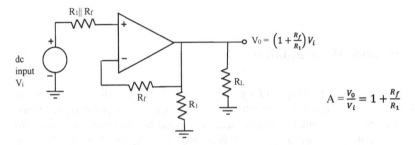

$$A = \frac{V_0}{V_i} = 1 + \frac{R_f}{R_1}$$

(b): Non-Inverting dc amplifier

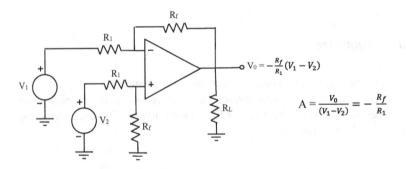

$$A = \frac{V_0}{(V_1 - V_2)} = -\frac{R_f}{R_1}$$

(c): Differential amplifier

Fig. 3.12 a Inverting DC amplifier, b non-inverting DC amplifier, c differential amplifier

$$f_L = \frac{1}{2\pi \left(R_{if} + R_0 \right) C_i}$$

where f_L is the lower cut-off frequency

    C_i is coupling capacitor or DC blocking capacitor

    R_{if} is AC input resistance of second stage and

    R_0 is the output resistance of first stage.

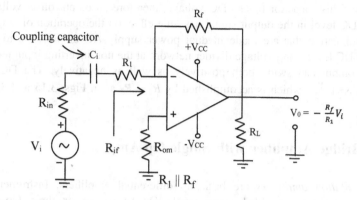

Fig. 3.13 AC inverting amplifier

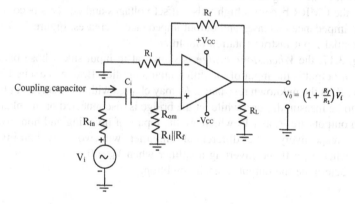

Fig. 3.14 AC non-inverting amplifier

The bandwidth of the amplifier is given by BW = $(f_H - f_L)$.

Where f_H is the high-frequency cut-off that depends on the closed-loop gain of the amplifier.

The value of f_H and f_L can be decided by the value of capacitor C_i, thus bandwidth can be fixed by the designer. The closed-loop gain of the amplifier in AC inverting mode is $A_f = -\frac{R_f}{R_1}$ and in AC non-inverting mode is $A_f = \left(1 + \frac{R_f}{R_1}\right)$.

3.5.3 AC Amplifier with a Signal Supply Voltage

In Op-Amp AC amplifier the single power supply can be used instead of ± 15 V dual power supply. This can be done by introducing an additional coupling capacitor. This capacitor blocks DC voltage, therefore an output offset voltage as well as DC level in the output signal has little effect on the operation of amplifier. The modification that are made in dual power supply AC amplifier is to insert positive DC level using voltage divider network at the non-inverting input terminal so that output can swing both positively as well as negatively. The DC level inserted is $+\frac{V_{cc}}{2}$, which is accomplished by $R_2 = R_3$ as in Figs. 3.15 and 3.16.

3.6 Bridge Amplifier with Single Op-Amp

Instrumentation amplifiers are basically differential amplifiers. Instrumentation amplifier can be assembled using single Op-Amp, two or three Op-Amps. Differential amplifier amplifies the difference between the two input signals. It is a combination of inverting and non-inverting amplifier. It lowers the common mode gain thus the CMRR becomes high. Also offset voltages and currents become low. The input impedance increases and output impedance decreases. Figure 3.17 shows the differential input instrumentation amplifier.

In Fig. 3.17, the *Wheatstone bridge* is shown at in input side whose balanced/ unbalanced output is the input of the differential amplifier. Bridge is excited by DC voltage V. R_x is the unknown resistance. R_x may change with the change in physical quantity to be measured. For a while let the bridge is disconnected or in unbalanced condition outputs are V_a and V_b which are the inputs of inverting and non-inverting terminals, respectively. The differential amplifier will work as non-inverting amplifier when $V_a = 0$ and inverting amplifier when $V_b = 0$.

Let us determine the output in each condition.

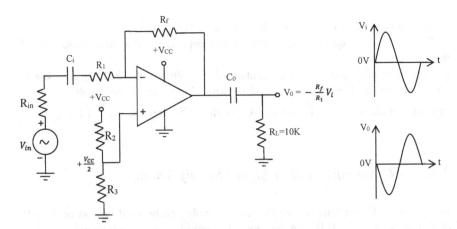

Fig. 3.15 AC inverting amplifier with single power supply

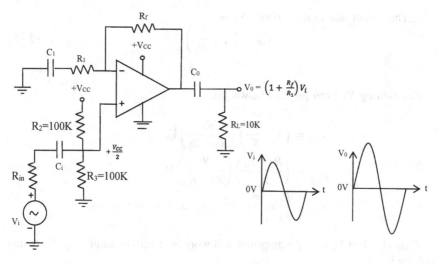

Fig. 3.16 AC non-inverting amplifier with single power supply

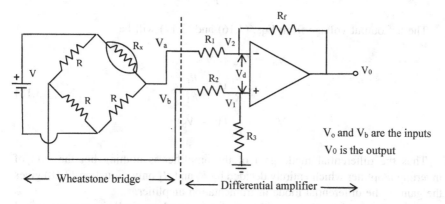

Fig. 3.17 Differential input instrumentation amplifier

Case I, when $V_a = 0$, the amplifier will work as non-inverting amplifier. The R_2 and R_3 form voltage divider network and voltage across R_3 is the voltage at non-inverting input V_1.

Therefore,

$$V_1 = \left(\frac{R_3}{R_2 + R_3}\right) V_b \tag{3.14}$$

$\therefore$ The output due to this voltage V_1 is

$$V_{ob} = \left(1 + \frac{R_f}{R_1}\right) V_1 \tag{3.15}$$

Substituting V_1 from Eq. (3.14), we get,

$$
\begin{aligned}
V_{ob} &= \left(1 + \frac{R_f}{R_1}\right)\left(\frac{R_3}{R_2 + R_3}\right) V_b \\
&= \left(\frac{R_1 + R_f}{R_1}\right)\left(\frac{R_3}{R_2 + R_3}\right) V_b \\
V_{ob} &= \frac{R_f}{R_1} V_b \quad (\because R_1 = R_2 \text{ and } R_3 = R_f)
\end{aligned} \tag{3.16}
$$

Case II, when $V_b = 0$, the amplifier will work as inverting amplifier. The output voltage is

$$V_{oa} = -\frac{R_f}{R_1} V_a \tag{3.17}$$

The net output voltage from Eqs. (3.16) and (3.17) will be

$$
\begin{aligned}
V_o &= V_{oa} + V_{ob} \\
&= -\frac{R_f}{R_1} V_a + \frac{R_f}{R_1} V_b \\
V_o &= -\frac{R_f}{R_1}(V_a - V_b)
\end{aligned} \tag{3.18}
$$

Thus the differential mode gain of the amplifier is nothing but the gain of inverting amplifier which entirely decided by R_f and R_1 only. Equation (3.18) gives the gain of the differential mode instrumentation amplifier.

As it will act as inverting amplifier, the input impedance is the input resistance of the amplifier.

$$\text{Hence } Z_{if_a} = R_1 \quad \text{when } V_b = 0 \tag{3.19}$$

The non-inverting mode impedance is

$$Z_{if_b} = (R_2 + R_3) \quad \text{when } V_a = 0 \tag{3.20}$$

However this will not affect the working much but the performance can be enhanced by modifying the circuit taking larger values of R_1 and $(R_2 + R_3)$ than source resistances.

3.7 Instrumentation Amplifier

Following Fig. 3.18 shows the simplified instrumentation amplifier using transducer bridge. A resistive transducer strain gauge is used. This transducer is connected in one arm of the bridge. The change in the resistance of the bridge is a function of change in physical quantity (strain/pressure). This bridge is excited by DC signal but it could be excited by AC signal also.

The bridge balanced condition is $V_M = V_N$

$$\text{or} \quad \frac{R_B(V_{ex})}{R_B + R_C} = \frac{R_A(V_{ex})}{R_A + R_T} \tag{3.21}$$

$$\text{i.e.} \quad \frac{R_C}{R_B} = \frac{R_T}{R_A} \tag{3.22}$$

The values of resistors R_A, R_B and R_C are selected so that they are equal in values to the transducer resistance R_T at some reference condition.

The reference condition is the specific value of physical quantity at which the bridge is balanced.

Initially bridge is balanced where $V_M = V_N$, but if physical quantity to be measured changes then bridge becomes unbalanced, i.e. $V_M \neq V_N$. Thus the output voltage of the bridge will change.

Let the change in resistance of the transducer be ΔR.

$$\text{Thus} \quad V_M = \frac{R_A}{R_A + (R_T + \Delta R)} V_{ex} \tag{3.23}$$

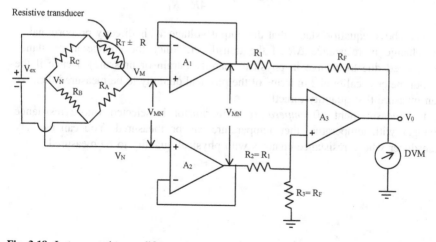

Fig. 3.18 Instrumentation amplifier

$$\text{and} \quad V_N = \frac{R_B}{R_B + R_C} V_{ex} \tag{3.24}$$

Hence the voltage V_{MN} will be $V_{MN} = V_M - V_N$ ($\because$ according to voltage divider rule at V_M and V_N points)

$$V_{MN} = \frac{R_A V_{ex}}{R_A + (R_T + \Delta R)} - \frac{R_B V_{ex}}{R_B + R_C}$$

If $R_A = R_B = R_C = R_T = R$, then

$$V_{MN} = -\frac{\Delta R V_{ex}}{2(2R + \Delta R)} \tag{3.25}$$

$-$ve sign indicates that $V_M < V_N$, because of increase in value of R_T.

This voltage V_{MN} is applied to the *differential instrumentation amplifier* composed of three Op-Amps A_1, A_2 and A_3. Op-Amps A_1 and A_2 act as *voltage followers* to avoid the loading of bridge.

So the gain of differential amplifier A_3 is $\left(-\frac{R_F}{R_1}\right)$, thus output V_0 would be

$$\begin{aligned} V_0 &= V_{MN}\left(-\frac{R_F}{R_1}\right) \\ &= \frac{\Delta R(V_{ex})}{2(2R + \Delta R)} \frac{R_F}{R_1} \end{aligned} \tag{3.26}$$

Generally the resistance of the transducer is very small so $(2R + \Delta R) \approx 2R$

$$\text{Hence} \quad V_0 = \frac{\Delta R V_{ex}}{4R} \frac{R_F}{R_1} \tag{3.27}$$

The above equation states that the output voltage V_0 is directly proportional to the change in resistance ΔR of the transducer. Since the change in resistance ΔR caused due to change in physical quantity (strain or pressure). Hence if the output meter is calibrated in terms of the physical quantity to be measured then we can measure that quantity directly.

For measurement of *temperature* if thermistor is selected whose resistance changes with temperature then temperature can be measured. You can use any transducer whose resistance changes with physical quantity to be measured.

Chapter 4
Linear Circuits

Applications of Op-Amp as *linear devices* as current-to-voltage and voltage-to-current converter, voltage and current measurements, summing, scaling and subtractor, integrator/differentiator circuits and analog computation, etc., are included in this chapter. The data acquisition and signal processing are the main components of any instrumentation system. Both circuits employed analog-to-digital (A/D) and digital-to-analog (D/A) conversion. Such converter circuits utilize Op-Amp. Importance of A/D and D/A converters and various types of them are discussed. As we have discussed Op-Amp is a multitasking and multidimensional versatile device thus it can be used in nonlinear functional circuits also.

4.1 Introduction

The operational amplifiers are used as variety of applications. The mode of operation of the amplifier can be applied to the performance as linear and nonlinear circuits. Since in inverting and non-inverting amplifiers the output is directly proportional to input, i.e. the Op-Amp operates in a linear mode. Therefore it can be classified as (a) linear circuits and (b) nonlinear circuits.

When output of the amplifier changes linearly with input called *linear circuits*. For example inverting, non-inverting amplifiers, integrators, differentiators, current-to-voltage converters, voltage-to-current converters, voltage followers, instrumentation amplifiers etc.

When output of the Op-Amp varies nonlinearly with input is called *nonlinear circuits*. The switching mode operation includes in a nonlinear operation of Op-Amp, for example zero crossing detectors, voltage comparators, Schmitt trigger, multivibrators—astable, monostable, bistable, etc.

S. Yawale and S. Yawale, *Operational Amplifier*,
https://doi.org/10.1007/978-981-16-4185-5_4

4.2   Current-to-Voltage Converter

The inverting amplifier can be used as current-to-voltage converter with small modification. Following Fig. 4.1 shows the circuit of current-to-voltage converter.
The output of the closed-loop inverting amplifier is

$$V_0 = -\frac{R_f}{R_1} V_i \quad \left(\because A_f = -\frac{R_f}{R_1} \right) = -\frac{V_i}{R_1} R_f \tag{4.1}$$

But $\frac{V_i}{R_1} = I_i$

$$\therefore \quad V_0 = -I_i R_f \tag{4.2}$$

Equation (4.2) shows that the output voltage is directly proportional to input current I_i. In other words the circuit will works as *current-to-voltage converter*. The gain of the amplifier decides by feedback resistor, R_f, so gain is called sensitivity of converter. For high sensitivity R_f should be large enough but it has certain limitations. To construct the high-sensitivity current-to-voltage converter (CVC), T-feedback network can be employed as in Fig. 4.2.

Fig. 4.1 Current-to-voltage converter

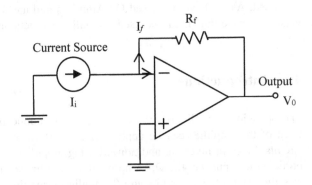

Fig. 4.2 CVC using T-network

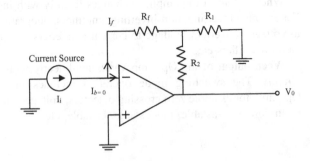

The output voltage is given by

$$V_0 = -\left(1 + \frac{R_2}{R_1} + \frac{R_2}{R_f}\right)I_i \qquad (4.3)$$

Equation (4.3) tells us that the output voltage is directly proportional to input current.

4.3 Voltage-to-Current Converter

In many instrument systems the input port is a physical quantity to be measured for which transducer can be used. The output of the transducer which converts physical quantity into electrical one is a weak voltage signal. This signal is not sufficient to drive the load, because it has very low current. In such cases the input analog voltage is necessary to convert into desired current which drives the next circuit or load. Such conversion of the signal is called voltage-to-current conversion. The device or circuit is called *voltage-to-current converter.*

There are two ways of conversion.

4.3.1 Floating Load Voltage-to-Current Converter

Figure 4.3 shows the floating load voltage-to-current converter.

The circuit is a simply current series negative feedback or non-inverting amplifier, where load resistor is floating, i.e. not connected to ground. Applying Kirchhoff's voltage law at the input side, we get

$$V_i = V_d + V_f \qquad (4.4)$$

But $V_d = 0$, since virtual ground or $A = \infty$
Therefore, $V_i = V_f$

Fig. 4.3 Floating load voltage-to-current converter

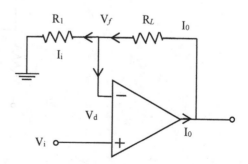

Or

$$V_i = I_i R_1 \quad (\because V_f = I_1 R_1)$$
$$V_i = I_0 R_1 \quad (\because I_1 = I_0)$$

(4.5)

Thus input voltage is directly proportional to the output current I_0.

4.3.2   *Grounded Load Voltage-to-Current Converter*

Another way to convert voltage to current is grounded load converter. In this circuit load current is controlled by input voltage. Non-inverting mode of Op-Amp is used.

Figure 4.4 shows the circuit of grounded load voltage-to-current converter. Applying Kirchhoff's current law at node N

$$I_L = I_1 + I_2$$

(4.6)

Calculate values of currents I_1 and I_2 and substitute in Eq. (4.6) to know voltage at node N say V_1

$$I_L = \frac{V_i - V_1}{R_1} + \frac{V_0 - V_1}{R_2}$$

Fig. 4.4 Grounded load
voltage-to-current converter

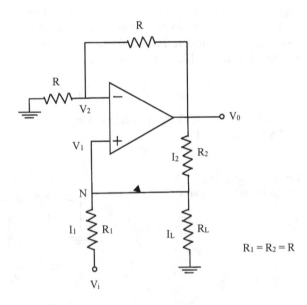

$$I_L = \frac{V_i - V_1 + V_0 - V_1}{R} \quad (\because R_1 = R_2 = R)$$
$$= \frac{V_i + V_0 - 2V_1}{R}$$

Or

$$V_1 = \frac{1}{2}(V_i + V_0 - I_L R) \tag{4.7}$$

The non-inverting mode is used so that gain of the amplifier is $1 + \frac{R_L}{R} = 2 = \frac{V_0}{V_1}$.
So

$$V_0 = 2V_1.$$

Substituting value of V_1 from Eq. (4.7), we get

$$V_0 = 2\left[\frac{1}{2}(V_i + V_0 - I_L R)\right]$$
$$V_0 = V_i + V_0 - I_L R$$

On rearranging the terms, we get

$$V_i = I_L R \tag{4.8}$$

This shows that the input voltage is directly proportional to load current I_L.

These circuits are used in AC/DC voltage and current measurements, LED and Zener diode testers, etc.

4.4 Voltage and Current Measurements

While measuring the quantity, voltage or current in the circuit should not affect the quantity being measured. In order to fulfil this, the voltmeter should have high input impedance and current meter should have low input impedance. Generally moving coil meters are used for measurement of these quantities which do not possess the above conditions. To meet or satisfy these conditions the Op-Amp is the best solution.

4.4.1 DC Voltage Measurement

As we know the voltage follower has very high input impedance and very low output impedance as well as unity gain, this circuit can be employed for measurement of DC voltage. Figure 4.5 shows the circuit of DC voltmeter.

Fig. 4.5 High input
impedance voltmeter

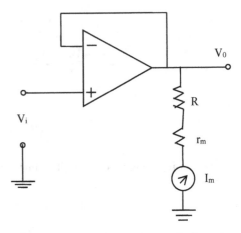

Resistance R is used for voltage scaling. The current flowing through meter is

$$I_m = \frac{V_0}{R + r_m} = \frac{V_i}{R + r_m} \quad \because V_0 = V_i(\text{unity gain}) \tag{4.9}$$

where r_m is the internal resistance of the meter.

Hence V_i is directly proportional to the current flowing through the meter. Calibrate the moving coil meter with standard input voltage, which gives linear relationship between standard and measured voltage values. The advantage of this voltage follower is that it provides high input impedance, hence avoids the loading cause due to moving coil meter.

4.4.2   DC Current Measurement

Op-Amp current-to-voltage converter can be employed for DC current measurement. Current meter using current-to-voltage converter is shown in Fig. 4.6.

Bridge T-network topography is employed in the circuit to enhance the sensitivity of the meter. This circuit provides very low input impedance as required for current meter. The output voltage is directly proportional to input current. Hence it can be used for current measurement. As we know in current-to-voltage converter the output voltage V_0 is

$$V_0 = -I_i R_f \tag{4.10}$$

The virtual ground is maintained at the point S, hence no current enters into the Op-Amp and entire current flows through the feedback resistor.

Fig. 4.6 Current meter using
Op-Amp

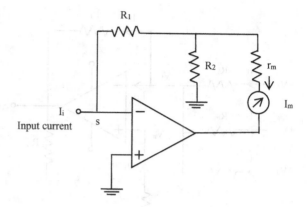

In both circuits voltage and current measurements are done with respect to ground terminal. Thus circuits are only useful for grounded measurements. For non-grounded measurements, i.e. in between the circuit or network it is not applicable. For such measurement differential input voltage amplifiers are to be employed.

4.5   Summing, Scaling and Averaging Amplifier

The *summing* (addition) action of Op-Amp can be explained by the addition of three voltages V_1, V_2 and V_3. We can add the numbers in terms of various voltages. Few external components say resistors are required to connect in the circuit. Following Fig. 4.7 shows the inverting configuration with three inputs V_1, V_2, and V_3. R_f is the feedback resistor. R_a, R_b and R_c are the input resistors.

This circuit is called *adder or summing amplifier, scaling amplifier or averaging amplifier*. Let us explain the working of this circuit by calculating output voltage.

Let us calculate the output voltage V_0. Applying Kirchhoff's current law at node V_b.

Thus we get

$$I_1 + I_2 + I_3 = I_F + I_{B2} \tag{4.11}$$

But $I_{B2} = 0$ and $I_{B1} = 0$ since R_i and A are ∞, i.e. $V_a = V_b = 0$ due to virtual ground.

Therefore,

$$\frac{V_1}{R_a} + \frac{V_2}{R_b} + \frac{V_3}{R_c} = \frac{V_b - V_0}{R_f}$$

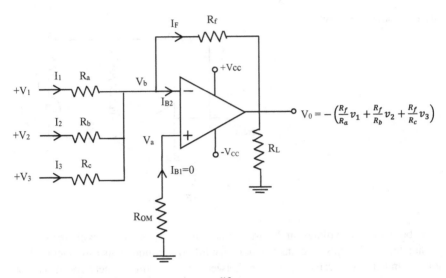

Fig. 4.7 Summing, scaling and averaging amplifier

$$\frac{V_1}{R_a} + \frac{V_2}{R_b} + \frac{V_3}{R_c} = -\frac{V_0}{R_F} \quad \because V_b = 0, I_1 = \frac{V_1}{R_a}, I_2 = \frac{V_2}{R_b}, I_3 = \frac{V_3}{R_c}$$

or

$$V_0 = -\left(\frac{R_f}{R_a}V_1 + \frac{R_f}{R_b}V_2 + \frac{R_f}{R_c}V_3\right) \tag{4.12}$$

If $R_a = R_b = R_c = R$, then

$$V_0 = -\frac{R_f}{R}(V_1 + V_2 + V_3) \tag{4.13}$$

Negative sign is due to inverting mode of the amplifier. From Eq. (4.13), we can say that the input voltages V_1, V_2 and V_3 are added. If $R_f = R$ then

$$V_0 = -(V_1 + V_2 + V_3) \tag{4.14}$$

Equation (4.14) shows the addition of three voltages.

Scaling amplifier

If $R_a \neq R_b \neq R_c$, i.e. R_a, R_b and R_c are different, i.e. $\frac{R_f}{R_a} \neq \frac{R_f}{R_b} \neq \frac{R_f}{R_c}$ means each input voltage is amplified by a different factor or weighted differently, then circuit is called scaling or weighted amplifier. Thus output of *scaling amplifier* is

$$V_0 = -\left(\frac{R_f}{R_a}V_1 + \frac{R_f}{R_b}V_2 + \frac{R_f}{R_c}V_3\right) \tag{4.15}$$

where, $\frac{R_f}{R_a} \neq \frac{R_f}{R_b} \neq \frac{R_f}{R_c}$.

Average Amplifier

In the adder circuit if $R_a = R_b = R_c = R$ then $\frac{R_f}{R} = \frac{1}{n}$, where n is the number of inputs.

Thus for three inputs $\frac{R_f}{R} = \frac{1}{3}$ ($\because n = 3$).
We get,

$$V_0 = -\left(\frac{V_1 + V_2 + V_3}{3}\right) \tag{4.16}$$

In all the cases the inputs V_1, V_2 or V_3 either could be AC or DC or mixed input voltages can be added.

4.6   Subtraction Circuit

Figure 4.8 shows the circuit of two input subtractors. Basically it is a scaling amplifier used in differential configuration.

The external resistors including feedback resistor all are having same value so the gain of the amplifier is equal to unity, i.e. 1.

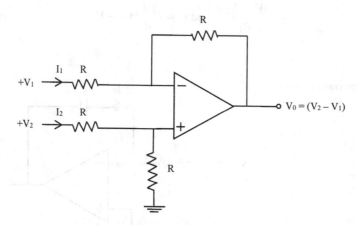

Fig. 4.8 Subtraction circuit

$$\therefore V_0 = -\frac{R}{R}(V_1 - V_2)$$

$$V_0 = (V_2 - V_1)$$

(4.17)

Thus the output voltage V_0 is equal to the subtraction of the voltages V_2 and V_1. Hence the circuit is called *subtractor*.

4.7 Integrator Circuit

In the basic inverting amplifier, if feedback resistance R_f is replaced by capacitor, then circuit works as integrator (Fig. 4.9).

The output can be obtained by applying Kirchhoff's current law at node V_2. So input current i_1 is given by

$$i_1 = i_B + i_f$$

(4.18)

Since $i_B \approx 0$ due to virtual ground and $A = \infty$ (ideal case)

$$\therefore i_1 = i_f$$

We know that current flowing through capacitor is given by

$$i_c = C\frac{dV_c}{dt}$$

($\because$ C is the value of capacitor and V_c is the voltage across capacitor)

$$i_1 = \frac{V_i - V_2}{R_1} \quad \text{and } i_f = C\frac{d}{dt}(V_2 - V_0)$$

(4.19)

Fig. 4.9 Integrator circuit

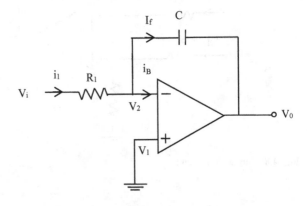

But $V_1 = V_2 = 0$, because $A = \infty$, then

$$\frac{V_i}{R_1} = C\frac{d}{dt}(-V_0)$$

Integrating both sides with respect to t

$$\int_0^t \frac{V_i}{R_1}dt = \int_0^t C\frac{d}{dt}(-V_0)dt$$

$$\int_0^t \frac{V_i}{R_1}dt = C(-V_0) + V_0 \mid_{t=0}$$

$$V_0 = -\frac{1}{R_1C}\int_0^t V_i dt + k \tag{4.20}$$

where k is the integration constant.

From above equation we can say that the output voltage V_0 is directly proportional to time integral of input signal and inversely proportional to time constant R_1C. Two types of voltages we can applied to the integrator as input such as DC and AC. Thus two cases may arise (Fig. 4.10).

(a) *If input voltage of the integrator V_i is constant, i.e. DC, then,*

$$V_0 = -\frac{1}{R_1C}\int_0^t V_i dt$$
$$= -\frac{V_i}{R_1C}t \tag{4.21}$$

Equation (4.22) indicates that the output of the integrator is directly proportional to time t as V_i R_1 and C are constant, i.e. the output V_0 increases with time t. Thus it ramp up and down as per the +ve and −ve DC voltage. The when square wave is applied at the input of the integrator then ramp up and down, i.e. triangular wave is generated at the output.

If the input to the integrator is a sinusoidal wave, i.e. $V_i = V \sin\omega t$, then

$$V_0 = -\frac{1}{R_1C}\int_0^t V\sin\omega t\, dt = \frac{1}{R_1C}\frac{V\cos\omega t}{\omega} \tag{4.23}$$

where ω is the frequency of input signal. The R_1 and C values are constant.

So $V_0 \propto \frac{1}{\omega}$, i.e. output is inversely proportional to frequency ω (Fig. 4.11).

Fig. 4.10 Waveforms

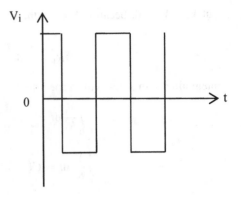

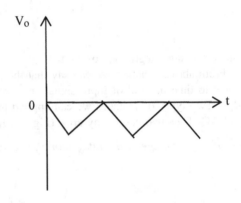

Fig. 4.11 Variation of output voltage (V_0) with frequency (f)

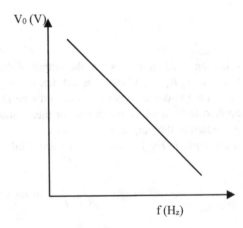

In this way we can study the integration by measuring the amplitude of the output signal at different frequencies.

4.8   Differentiator Circuit

The *differentiator* circuit can be prepared by interchanging the positions of capacitor and input resistor R_1. In operational amplifier differentiator basic circuit is inverting amplifier. Figure 4.12 shows the differentiator circuit using operational amplifier.

Applying Kirchhoff's current law at node V_2,

$$\therefore i_C = i_B + i_f \tag{4.24}$$

Since $i_B = 0$ because $V_1 = V_2 = 0$ and $A = \infty$, $R_i = \infty$ then

$$i_C = i_f$$

$$C\frac{d}{dt}(V_i - V_2) = \frac{V_2 - V_0}{R_f}$$

$\therefore i_C = C\frac{d}{dt}(V_i - V_2)$ $\because (V_i - V_2)$ is the voltage across capacitor C.
And

$$i_f = \frac{V_2 - V_0}{R_f}.$$

But $V_1 = V_2 = 0$ as $A = \infty$

Fig. 4.12 Differentiator circuit

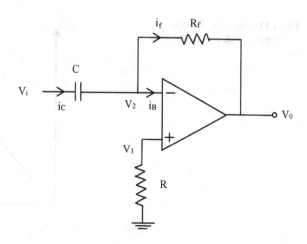

$$C \frac{d}{dt} V_i = -\frac{V_0}{R_f}$$

$$V_0 = -CR_f \frac{d}{dt} V_i \tag{4.25}$$

where C and R_f are constants having fixed value. Thus output voltage V_0 is the time differential of input signal V_i. Since the differentiator performs the reverse of the integrator function, a cosine wave input will produce sinusoidal wave output and triangular wave will produce a square wave output.

If a sinusoidal voltage is applied as input then $V_i = V \sin \omega t$.

Hence

$$V_0 = -R_1 C \frac{dV_i}{dt} = -R_1 C \frac{d}{dt} (V \sin \omega t)$$

$$V_0 = -R_1 C \, \omega V \cos \omega t \tag{4.26}$$

where R_1 and C values are constant.

Thus output V_0 is proportional to frequency ω (Fig. 4.13). In other words the amplitude of the output signal increases with input frequency signal.

We can select proper values of R_1 and C as $R_1 = 1$ M and $C = 1$ μF then

$$R_1 C = 1$$

$$\therefore V_0 = -\frac{d}{dt} V_i \tag{4.27}$$

This gives direct differentiation of input signal.

Fig. 4.13 Variation of output voltage (V_0) with frequency (f)

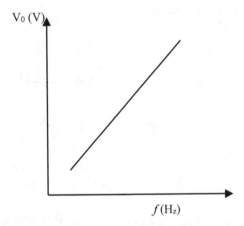

4.9   Analog Computation

Analog computer is used to solve the differential and simultaneous equations. It can solve various types of problems in mathematics. The system with few components such as resistors, capacitors and operational amplifiers can be designed to solve the mathematical problem. It is applicable in solving network analysis and transfer function problems. Especially in electrical networks or systems or circuits and mechanical engineering these are useful.

4.9.1   Solution to the Simultaneous Equations

The simultaneous equations can be solved by using Op-Amp analog computer.

Let us have the following two equations having variables x and y. Solving these equations simultaneously we can obtain values of x and y.

$$x + 4y = 3 \qquad (4.28)$$

and

$$2x - 5y = 6 \qquad (4.29)$$

To design a circuit using Op-Amp for evaluation of x and y variables.
On rearranging these equations we have

$$x = 3 - 4y \qquad (4.30)$$

and

$$y = -\frac{6}{5} + \frac{2}{5}x \qquad (4.31)$$

Looking to Eqs. (4.30) and (4.31) we require two adders to get the values of x and y. In Eq. (4.30), 3 and $-4y$ are added for obtaining value of x and in Eq. (4.31), $-\frac{6}{5}$ and $\frac{2}{5}x$ are added to get the value of y.

In the circuit Op-Amp A is working as adder whose inputs are $+4y$ and -3 V. -3 V is taped from -5 volt power supply and $+4y$ voltage is generated as output. In adder amplifier all resistances are having same value as R. On addition of $+4y$ and -3 V the output generated is x. Thus on addition we get

$$V_0 = -(4y - 3) = -4y + 3 = x \qquad (4.33)$$

As per Eq. (4.31) we have to add −6/5 and (2/5)x. This x output is tapped by the potentiometer to get $\frac{2}{5}x$ voltage. This voltage is given to sign changer to get $-\frac{2}{5}x$ Value. This work is done by Op-Amp B in the circuit.

Op-Amp C is working as adder to get the value of y. The inputs are $-\frac{2}{5}x$ and $-\frac{6}{5}V$. Thus output of Op-Amp C is

$$
\begin{aligned}
V_0 &= -\left(-\frac{2}{5}x - \frac{6}{5}\right)\\
&= \frac{2}{5}x + \frac{6}{5} = y
\end{aligned}
\tag{4.34}
$$

This $-\frac{6}{5}V$ is tapped through potentiometer.

The Op-Amp D is working as inverting amplifier having gain 4. The output of this amplifier is −4y. To get the output of +4y unity gain inverter is used. This gives the output +4y which is one of the inputs of the Op-Amp A. In this way the corresponding voltages are generated (Fig. 4.14). The values of x and y variables are generated in the form of voltages.

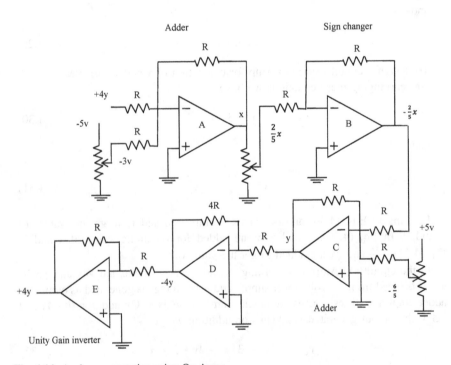

Fig. 4.14 Analog computation using Op-Amps

4.9.2 Differential Equations

Consider the following differential equation.

$$\frac{d^2x}{dt^2} + K_1 \frac{dx}{dt} + K_2 x - y = 0 \qquad (4.35)$$

where x is the function of time and K_1 and K_2 are the constants.

Let initially the $\frac{d^2x}{dt^2}$ is available in the form of voltage. To obtain $\frac{dx}{dt}$ and x, two integrators are required.

The output of integrator A is $\frac{dx}{dt}$ which forms the input of integrator B. So the output of integrator B will be x. Tapping the x signal as $K_2 x$ and $\frac{dx}{dt}$ as $K_1 \frac{dx}{dt}$, added with $-y$ gives the output of Op-Amp C as $\frac{dx}{dt}$. The Op-Amp C will work as adder. Therefore the output of adder will be

$$V_0 = -\left(K_1 \frac{dx}{dt} + K_2 x - y \right)$$
$$= \frac{d^2x}{dt^2} \qquad (4.36)$$

Hence the input is regenerated. This signal or the output of adder is connected as input of Op-Amp A integrator will solve the differential equation.

Care must be taken during the integration of Op-Amps A and B that initial conditions or voltages are proportional to $\frac{dx}{dt}$ and x values. In adder circuit it is necessary to apply all voltages simultaneously. So the voltage y must be applied through switch. The voltages generated at the outputs of Op-Amps A and B are proportional to the values of $\frac{dx}{dt}$ and x (Fig. 4.15).

4.10 Digital-to-Analog (D/A) and Analog-to-Digital (A/D) Converters

4.10.1 Introduction

Most of the data in a real world is available in analog form. For example temperature, pressure, velocity, acceleration, light, intensity, etc., are continuous signals and are linear. This data or signal could not be processed in digital circuits, i.e. microprocessor or computer. To manipulate this analog data using digital system or device, it is necessary to convert the analog signals-to-digital signals so that the digital device will be able to read, understand and manipulate the signal. Not only

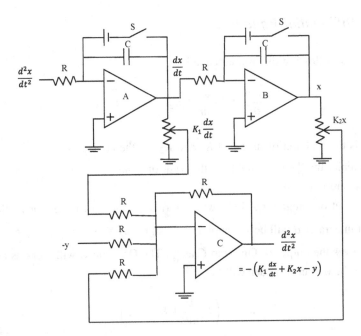

Fig. 4.15 Solving differential equation

this but after processing analog data by the digital device then it should be converted back to readable format. Thus it requires digital signal to be converted into analog signal. ADC/DAC are used in music players, mobiles, instrumentation systems, televisions, sound cards, digital speakers, signal communications, optical communications, etc.

4.10.2   Digital-to-Analog (D/A) Converter

There are two types of D/A converters such as binary weighted resistors and binary ladder. Both types are very popular and hence discussed here.

4.10.3   (a) Binary Weighted Resistors Type

After processing the analog signal, the binary output is available in the form of 0s and 1s called as digital output. To convert this digital output in the form of analog signal the device used is called *digital-to-analog (D/A) converter*. Following Fig. 4.16 shows the binary weighted resistor-type D/A converter. It is simply an inverting Op-Amp adder circuit.

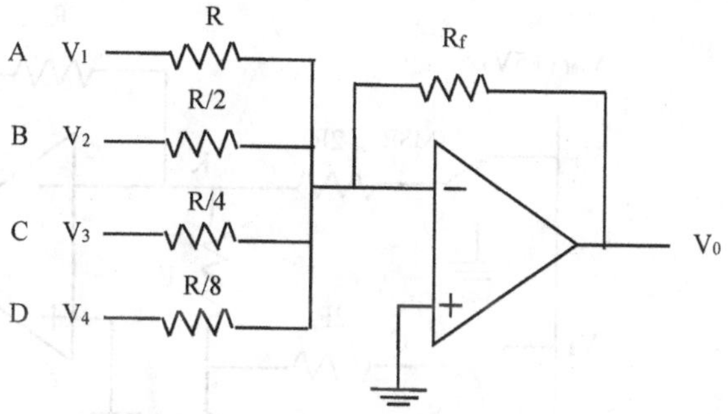

Fig. 4.16 4-bit binary weighted resistor-type D/A converter

The digital inputs applied in the form of +5 V and 0 V as '1' and '0' level in digital system, respectively. The values of resistors connected in the circuit have values in the ratio of respective weight positions in 4-bit binary numbers as 2^4, 2^2, 2^1 and 2^0, i.e. 8, 4, 2 and 1. Let V_1, V_2, V_3, and V_4 are the inputs of adder. Input voltage may be either +5 V or 0 V depending on binary input available. The output of adder is given by

$$V_0 = -R_f \left(\frac{V_1}{R} + \frac{V_2}{R/2} + \frac{V_3}{R/4} + \frac{V_4}{R/8} \right) \tag{4.37}$$

If $R_f = R$, then

$$V_0 = -(V_1 + 2V_2 + 4V_3 + 8V_4) \tag{4.38}$$

The proper value of R_f can be selected to get desired value of output voltage step. The value of R_f should be chosen properly so that the output voltage of Op-Amp does not exceed the saturation value of Op-Amp. In the conversion of larger bit binary number, the large numbers of resistors of related positional values are required. If accurate value of resistor is not available the output becomes erroneous. To overcome these difficulties binary ladder or R-2R type converters are used.

4.10.4 (B) Binary Ladder or R-2R Type Converter

A circuit utilizing binary ladder consisting of identical resistors R and 2R is shown in Fig. 4.17.

In Fig. 4.17 S_1, S_2, S_3 and S_4 are the switches. The position of electronic switch may change as per the input. For '1' it is connected to +5 V, V_{ref} and for '0' it is connected to ground terminal.

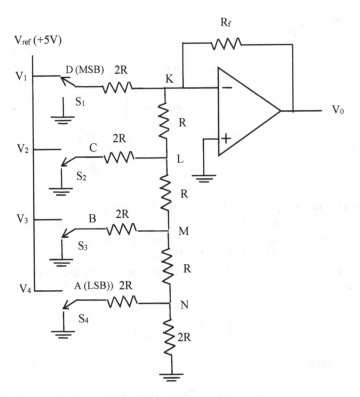

Fig. 4.17 Binary ladder or R-2R type converter

Consider the binary input D, C, B, A having positional weights at respective bit positions 8, 4, 2, 1. Let DCBA—1000, i.e. D is connected to V_{ref} and others C, B and A to ground.

Applying Thevenin's theorem from bottom to the R-2R ladder or network, the resistance

$(2R \parallel 2R) + R = 2R$—at point M
$(2R \parallel 2R) + R = 2R$—at point L
$(2R \parallel 2R) + R = 2R$—at point K.

Finally the network will be looked as follows (Fig. 4.18).

Now let DCBA = 0100, proceeding on the similar lines, the output voltage at point K would be

$$V_0' = \left(\frac{2R}{R + R + 2R} \right) \frac{V_{ref}}{2} \quad \left(\because V_{ref} \text{ is } \frac{V}{2} \right)$$

$$V_0' = \left(\frac{1}{4} \right) V_{ref} \tag{4.39}$$

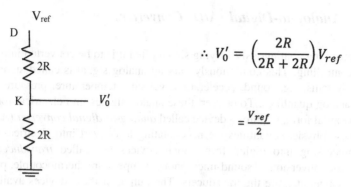

$$\therefore V_0' = \left(\frac{2R}{2R + 2R}\right) V_{ref}$$

$$= \frac{V_{ref}}{2}$$

Fig. 4.18 Network

Thus when DCBA = 0010, the output voltage V_0' will be $\left(\frac{1}{8}\right) V_{ref}$ and so on. By using the principle of superposition if '1' appears in a four number of places the output V_0 will be

$$V_0 = -R_f \left(\frac{V_1}{2R} + \frac{V_2}{4R} + \frac{V_3}{8R} + \frac{V_4}{16R}\right) \tag{4.40}$$

If $R_f = R$ and n-bit ladder is connected, then

$$V_0 = -\left(\frac{V_1}{2R} + \frac{V_2}{4R} + \frac{V_3}{8R} + \frac{V_4}{16R} + \cdots + \frac{V_n}{2^n}\right) \tag{4.41}$$

The selection of D/A converter depends on its specifications which include resolution, accuracy, linearity and settling time.

(a) *Resolution*

It is the smallest change in the output analog voltage produced. It depends on the number of input bits. The 4-bit D/A converter has resolution one part in $(2^4 - 1)$, i.e. 15.

(b) *Accuracy*

Accuracy is the deviation of output voltage with ideal one. It is expressed as percentage of maximum output voltage. Ideally the accuracy should be $\pm\frac{1}{2}$ LSB.

(c) *Linearity*

It is the precision with which the digital input is converted into analog output. Linearity error is the deviation from the ideal step size of the D/A converter.

(d) *Settling time*

The time is required by D/A converter to settle output with in $\pm\frac{1}{2}$ LSB of its final value when change occurs at the input code.

4.10.5   Analog-to-Digital (A/D) Converters

While processing continuously varying signal often it is to be converted into digital one for controlling. This continuously varying analog signal is called signal from physical systems, e.g. sound, acceleration, velocity, temperature, pressure, stress, etc., are analog quantities. To convert these analog signals into electrical form and then into digital form, requires a device called *analog-to-digital converter (A/D)*. Of course these physical quantities are necessitating to convert into electrical signals before processing into digital form. Such devices are called *transducers*. For example, for conversion of sound-microphone, temperature-thermocouple, pressure —strain gauge, etc., are the transducers. The output of these devices available in electrical form can be converted into a digital form.

There are different types of A/D converters; few of them are listed below.

(a) Counter-type A/D converter; (b) parallel comparable or flash A/D converter (c) voltage-to-time A/D converter (d) dual slope A/D converter; (e) successive approximation A/D converter.

In this section we shall discuss only counter type and successive approximation A/D converters.

4.10.6   (a) Counter-Type A/D Converter

The block diagram of counter-type A/D converter is shown in Fig. 4.19.

Initially a start pulse is applied to monostable multivibrator and the same pulse simultaneously applied to reset input of N-bit binary counter as shown in Fig. 4.19.

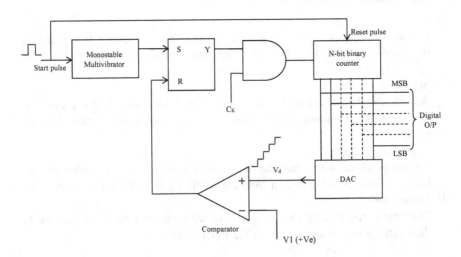

Fig. 4.19 Block diagram of counter-type A/D converter

The output of the monostable multivibrator is connected to set (S) input of SR flip-flop. The monostable multivibrator introduces delay between setting of y output of SR flip-flop and resetting of N-bit binary counter. Now the binary counter is ready for counting the pulse coming from AND gate. This time output of counter is zero hence the output of D/A converter (DAC) is also zero. This output is provided to the non-inverting input of comparator. The voltage V_1 (+ve), i.e. analog voltage to be converted is applied to the inverting input of the comparator. Since it is +ve, the output of the comparator goes low which makes $R = 0$ of SR flip-flop i.e. *SR* flip-flop resets. But $S = 1$ because start pulse is applied through monostable multivibrator; hence it will set the SR flip-flop, i.e. output $y = 1$. So one input of AND gate is high and clock pulses are applied thus AND gate becomes enabling and the clock pulses reach to the clock of N-bit binary counter that increases and simultaneously the binary signal is converted into analog signal in steps. This analog signal V_d compares with the analog signal V_1 (to be converted into equivalent digital). As soon as both the analog signals (V_d and V_1) becomes equal, the output of the comparator changes from low to high. Hence $R = 1$, while $S = 0$, which makes the output of the SR flip-flop to zero, i.e. $y = 0$, the AND gate is disabled and it stops reaching the clock pulses to the N-bit binary counter. The counter stops and the binary output of the counter is equivalent of the V_1 analog voltage.

The time diagram is shown in Fig. 4.20.

Drawbacks

(1) It requires longer time for conversion.
(2) The clock pulse rate decides the conversion time.
(3) The conversion time is different for voltages of different magnitude.
(4) It is not possible to convert the analog signal continuously for an analog voltage varies with time.

4.10.7 (b) Tracking or Servo Type

The conversion speed is important factor in ADC. This will decide the selection of ADC. An improved version of counter-type ADC utilizes up/down counter is called a tracking or servo or continuous type A/D converter. The modified version of counter type ADC is shown in Fig. 4.21.

In this circuit neither a start pulse or reset pulse nor an AND gate is necessary. Only up/down counter is required and comparator output can be used as control signal for logic circuit to control the up and down counting of the counter.

As soon as the supply to the counter is on it will have some count produces analog equivalent of the count by the DAC. This signal compares with the V_1 (analog signal to be converted). If this is less than V_1 then positive output of the comparator causes the counter to count up. Thus the binary output of the up/down counter goes on increasing until it exceeds V_1. As it crosses V1 the output of the

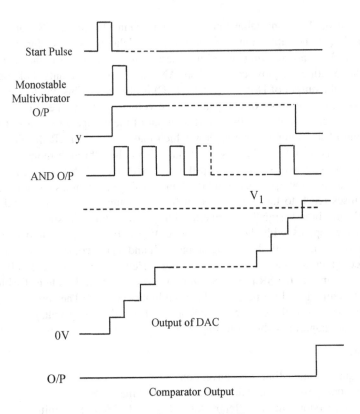

Fig. 4.20 Time diagram of counter-type A/D converter (ADC)

comparator changes its state so that the counter starts down counting (by 1 LSB). This process of up and down counting by 1 LSB keeps repeating so the binary output becomes back and forth by ±1 LSB or the correct value.

4.10.8 (c) Successive Approximation A/D Converter

Figure 4.22 shows a successive approximation ADC. The main component in the circuit is successive approximation register (SAR). In some circuits programmer is used. The output of the SAR is applied to the digital-to-analog converter (DAC).

The output obtained from DAC is compared with the analog voltage (V_1) to be converted. This output of the comparator is a serial data input to the SAR. This SAR adjusts the output till the equivalent input voltage is obtained.

Working

Applying the start pulse the SAR resets. At the first clock pulse the MSB of the SAR sets, i.e. $Q_7 = 1$ (for 8-bit). Thus output of SAR will be 10,000,000.

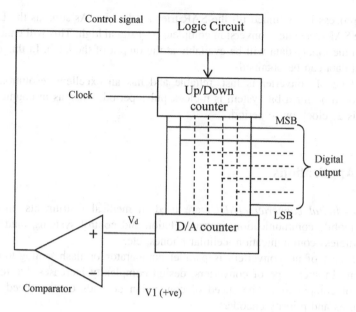

Fig. 4.21 Modified version

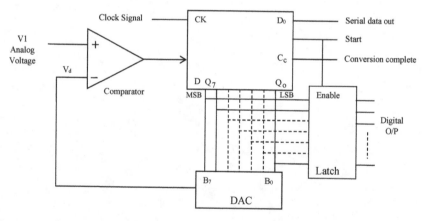

Fig. 4.22 Successive approximation ADC

Then DAC generates the analog equivalent voltage of Q_7 bit which is compared with the V_1, i.e. V_d compared with V_1. If the comparator output is low, the DAC output $V_d > V_1$ and SAR will clear its MSB Q_7. On the other hand, if the comparator output is high, the DAC output $V_d < V_1$ and SAR will keep the MSB Q_7 to set.

At the next clock pulse, the SAR will set Q_6, depending on output of the comparator the bit of SAR will set or reset.

This process is continued till the SAR tries all the bits. As soon as the LSB is tried, the SAR generates conversion complete (C_c) signal high. This will enable the latch and the binary data will be available at the output of the latch. In this circuit the serial data can be obtained.

This type of converter is fast, reliable and has an excellent resolution. The conversion time for n-bit system is n-clock pulse period, whereas in counter type ADC it is 2^n clock pulse period.

4.11   Applications

Analog-to-digital converters (ADC) are used in medical instruments, modems, routers, mobile communication, instrumentation and control systems, data acquisition, wireless communication, cellular phones, etc.

The fastest of all converters is parallel comparator or flash analog-to-digital converters. In such type of converters, design complexity increases due to more number of comparators. The speed of conversion depends on the speed of the comparators and priority encoders.

Chapter 5
Nonlinear Circuits

As discussed in Chap. 4 the working of operational amplifier as linear circuits, this chapter contains the discussion on logarithmic and exponential amplifiers, multiplier and amplitude modulator and so on. The operational amplifier is multi-dimensional and multitasking amplifier, and hence it can be utilized for design of any circuits. We are all aware of that the low amplitude or weak signals having millivolt strength could not be rectified by ordinary diode rectifier. Because the *cutin* voltage of the ordinary *p–n junction* diode is 0.7 V for Si diode. Unless the input voltage of the reciter is more than this it will not be possible to get the rectified signal at the output. In such cases the peak detectors and precision rectifiers play prominent role. These precision rectifiers are made up of operational amplifiers. The concept of half and full wave precision rectifiers and peak detectors is described. In a data acquisition system sample and hold circuit have prime importance. Hence the concept of this circuit is also discussed.

5.1 Introduction

It is very well known that operational amplifier works as a linear device. The linear operation of Op-Amp in integrator, differentiator, voltage to current, current to voltage converters, etc., is described and explained.

In an electronic circuit, when relationship between input and output is nonlinear then the electronics circuit is called nonlinear circuit. In this chapter, the application of Op-Amp as a nonlinear device is discussed. The nonlinear circuits such as precision rectifiers, peak detectors, logarithmic amplifier, anti-logarithmic amplifier and multiplier are discussed. These nonlinear circuits are mostly used in analog computing circuits to generate nonlinear signal.

© The Author(s), under exclusive license to Springer Nature Singapore Pte Ltd. 2022
S. Yawale and S. Yawale, *Operational Amplifier*,
https://doi.org/10.1007/978-981-16-4185-5_5

5.2 Precision Rectifiers

The device used for conversion of alternating current to direct current is called rectifiers, and the process is called rectification. Generally we use diodes as rectifiers. While using diode as rectifier the input AC voltage applied to the diode should be greater than 0.7 V, called cut in or forward voltage. For lower voltages than forward diode voltage rectification is not possible. The rectifiers which rectify the low value voltages of the value of mV using Op-Amp is known as precision rectifiers.

As usual there are half wave and full wave rectifiers.

5.2.1 Half Wave Precision Rectifier

Figure 5.1 shows the half wave precision rectifier in which rectifier diode D is connected at the output. The non-inverting configuration of operational amplifier is used. The effective output of the circuit is feedback through inverting input and input is directly applied through non-inverting input of Op-Amp.

During positive half cycle, the diode D conducts transferring this half cycle at the output; while in negative half cycle of input, diode D becomes reverse biased, and thus the negative half cycle prevents to reach at the output. Because of high open loop gain of the amplifier the voltage drop across the diode gets adjusted, and it will behave like an ideal diode. Hence the rectified output voltage is exactly equal to the AC input voltage. The input and output waveforms are shown in Fig. 5.2.

For the rectification of negative half cycle just reverse the polarity of diode connected at output in positive half wave for precision rectifier. The slew rate and response time are important factors. This may limit the frequency range at the input. However this can be rid of using high speed of Op-Amps such as LM310 and μA 318. The undershoot problem creates the spikes at the output when the Op-Amp goes into negative saturation. The high slew rate amplifiers can solve this problem. The modified circuit of half wave precision rectifier which provides finite gain and prevents the saturation of the Op-Amp is shown in Fig. 5.3a, b.

Fig. 5.1 Half wave precision rectifier

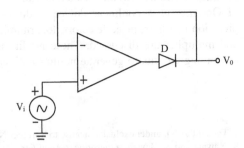

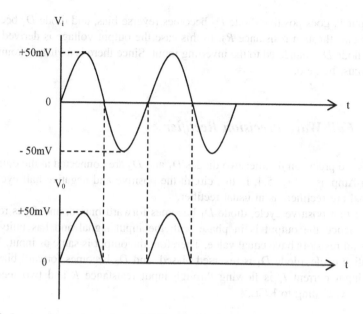

Fig. 5.2 Input and output waveform of half wave rectifier

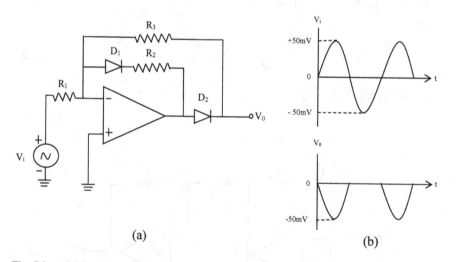

Fig. 5.3 a, b Modified circuit of half wave precision rectifier with waveforms

The inverting mode of the amplifier is used in the circuit. If input V_i goes negative, the output of the Op-Amp will be positive that makes diode D_2 forward bias and diode D_1 reverse bias. In this case the circuit operates as inverting amplifier having gain R_3/R_1.

If input V_i goes positive diode D_2 becomes reverse bias, and diode D_1 becomes forward bias through resistance R_2. In this case the output voltage is derived from R_3 and diode D_1 connected to the inverting input. Since there is a virtual ground the output must be zero.

5.2.2 Full Wave Precision Rectifier

In full wave precision rectifier two diodes D_1 and D_2 are connected at the output of first Op-Amp as in Fig. 5.4. In this circuit the positive and negative half cycles of the signal are rectified as in usual rectifier.

In the first positive cycle, diode D_1 becomes forward biased, and D_2 is reverse biased. Hence the output is in phase with the input signal and has unity gain because all resistors have equal value. Therefore the output is same as input. While in negative cycle, diode D_1 is reversed biased, and D_2 becomes forward biased.

Let input current I_1 is flowing through input resistance R and two feedback resistors. According to KCL,

$$I_1 = I_2 + I_3 \tag{5.1}$$

But $I_1 = \frac{V_i}{R}$, $I_2 = \frac{V_0}{3R}$ and $I_3 = -\frac{V_2}{R}$.
So,

$$\frac{V_i}{R} = -\frac{V_0}{3R} - \frac{V_2}{R} \tag{5.2}$$

However, $V_2 = V_3 = \frac{2R}{3R} V_0$

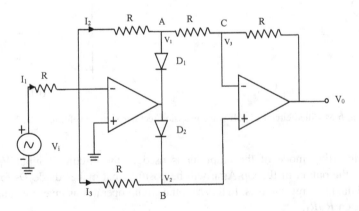

Fig. 5.4 Full wave precision rectifier

$$V_2 = \frac{2}{3}V_0$$

Substituting value of V_2 in Eq. (5.2), we get

$$\frac{V_i}{R} = -\frac{V_0}{3R} - \frac{2}{3R}V_0$$

On solving we get

$$V_i = -V_0 \tag{5.3}$$

or $V_0 = -V_i$.

5.3 Peak Detector

The circuit which performs to measure the maximum peak and negative valley value of a waveform over a specified time is called peak detector, or it is sometimes called as peak follower. It consists of diodes and capacitor. There are various types of peak detector circuits. Few of them are shown below. Figure 5.5a, b shows circuit of peak detector.

The non-sinusoidal signal is given through non-inverting input. When input signal V_i is positive, the positive output of the Op-Amp makes diode D_1 forward biased and transfers the output of Op-Amp to capacitor C. This capacitor C charges to this positive peak value of input V_i, since $V_0 = V_i$ because Op-Amp works as voltage follower. During negative half cycle, the diode D_1 is reversed biased, and D_2 conducts. The voltage across capacitor holds. The diode D_2 prevents the Op-Amp going to negative saturation because it becomes forward bias. The charging time of capacitor is R_dC, where R_d is the diode resistance in forward bias mode, and discharging time is R_LC, where R_L is the load resistance. The input and corresponding output waveforms are shown in Fig. 5.6.

By reversing the diode negative peaks of the input signal V_i can be detected. Such circuits are useful for AC magnitude measurement in non-sinusoidal waveforms. The traditional AC voltmeters measured *RMS* value of sin wave. In case of non-sinusoidal signals the peak values are being measured. The peak detector measures the positive peak value of square waveform.

In case of irregular waveform where amplitude is not constant, the circuit first tries to detect the available signals on a first-come basis, and the capacitor holds that signal. While the next pulse reaches and having same high peak value as that of earlier, then again the capacitor changes to that value and retains the change. So this will be the output voltage of peak detector finally.

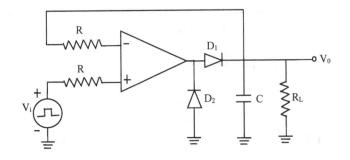

(a)

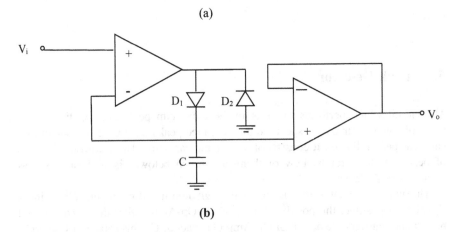

(b)

Fig. 5.5 a, b Peak detector circuit

Peak detectors are mostly used in AC measuring instruments especially AC voltmeters where square, saw tooth or any random waveform magnitude are not measured by usual AC voltmeters. It is used in mass spectrometers, AM Communication circuits, etc.

5.4 Logarithmic Amplifier

This amplifier produces the output proportional to the logarithm of the input voltage. In this case the Op-Amp operates in a nonlinear mode. A diode is used as a feedback element in the inverting Op-Amp circuit. Figure 5.7 shows the circuit of logarithmic amplifier.

We know the nonlinear volt-ampere (V–I) relationship of p–n junction diode in a forward bias mode. The forward current flowing through the diode is given by

Fig. 5.6 Input and output
waveforms of peak detector

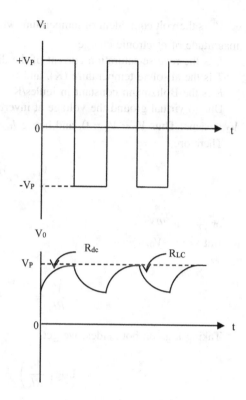

Fig. 5.7 Logarithmic
amplifier

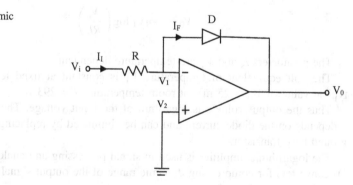

$$I_f = I_s \left(e^{V_f/nV_T} - 1 \right)$$

$$I_s = e^{V_f/nV_T} \quad \text{since } e^{V_f/nV_T} \gg 1.$$

(5.4)

where V_f is the voltage drop across the diode in forward bias mode

V_T is the volt equivalent of temperature which is given by $V_T = \frac{KT}{q}$, and q is the magnitude of electronic charge

I_s is the reverse saturation current of the diode

T is the absolute temperature (K), and

K is the Boltzmann constant in joules/K.

Due to virtual ground the voltage at inverting input V_1 and non-inverting input V_2 is same. Thus $V_1 = V_2 = 0$, and hence $I_1 = I_F = \frac{V_i}{R}$.

Therefore,

$$I_F = \frac{V_i}{R} = I_s e^{v_f/nV_T} \tag{5.5}$$

or $\frac{V_i}{RI_s} = e^{v_f/nV_T}$.

But $v_f = -V_0$.

so

$$\frac{V_i}{RI_s} = e^{-V_0/nV_T} \tag{5.6}$$

Taking logs on both sides, we get

$$\mathrm{Log}\left(\frac{V_i}{RI_s}\right) = -\frac{V_0}{nV_T}$$

$$V_0 = -nV_T \log\left(\frac{V_i}{RI_s}\right) \tag{5.7}$$

The parameters I_s and V_T are temperature-dependent.

The volt equivalent of temperature V_T is constant at fixed temperature. The typical value of $V_T \approx 25$ mV at room temperature $T = 293$ K.

Thus the output voltage is logarithm of the input voltage. The value of factor n depends on the diode current and can be eliminated by replacing the diode with ground base transistor.

The logarithmic amplifier is used in signal processing and analog to digital (A/D) converters for compressing dynamic range of the output signal.

5.5   Antilog or Exponential Amplifier

Antilog or exponential amplifier performs the inverse function of logarithmic amplifier. The output voltage is proportional to the base e or 10. In antilog amplifier, the position of diode and feedback resistor is interchanged as provided in logarithmic amplifier. Figure 5.8 shows the circuit of antilog or exponential amplifier.

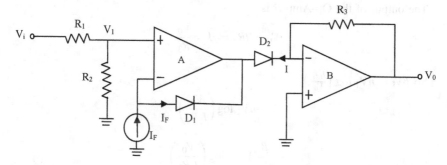

Fig. 5.8 Exponential or antilog amplifier

In above circuit two Op-Amps A and B are used. In first part diode D_1 is used as feedback element. The second part of the circuit R_3 utilizes as feedback element and D_2 as input element.

According to the voltage divider rule, the voltage V_1 at non-inverting input of Op-Amp A is

$$V_1 = \frac{R_2}{R_1 + R_2} V_i = \beta V_i \quad \left(\because \beta = \frac{R_2}{R_1 + R_2} \right) \tag{5.8}$$

The output voltage V_0' obtained at the Op-Amp A is

$$V_0' = V_1 - V_D$$

$$= \beta V_i - nV_T(\log I_f - \log I_s) \tag{5.9}$$

Since $V_D = nV_T(\log I_f - \log I_s)$.

Where I_s is the reverse saturation current of the diode, and V_T is the volt equivalent of temperature.

Also

$$V_0' = -nV_T(\log I - \log I_s) \tag{5.10}$$

where I is the current flowing through diode D_2.

From Eqs. (5.9) and (5.10), equating we get,

$$\beta V_i - nV_T(\log I_f - \log I_s) = -nV_T(\log I - \log I_s)$$

or $\beta V_i - nV_T \log I_f = -nV_T \log I$

or $\beta V_i = -nV_T \log I + nV_T \log I_f$

$$\beta V_i = nV_T \log \frac{I_f}{I} \tag{5.11}$$

The output of the Op-Amp B is

$$V_0 = IR_3 \Rightarrow I = \frac{V_0}{R_3}$$

So $\beta V_i = nV_T \log\left(\frac{I_f R_3}{V_0}\right)$

$$= -nV_T \log\left(\frac{V_0}{I_f R_3}\right)$$

$$-\frac{\beta}{nV_T} V_i = \log\left(\frac{V_0}{I_f R_3}\right)$$

or $\frac{V_0}{I_f R_3} = \exp\left(-\frac{\beta}{nV_T} V_i\right)$

$$V_0 = I_f R_3 \exp\left(-\frac{R_2}{R_1 + R_2}\frac{V_i}{nV_T}\right) \tag{5.12}$$

Thus the output V_0 is the exponential or antilogarithm of input voltage V_i. The parameter n depends on the kind of semiconductor used. It is approximately 2 for silicon and $V_T \approx 25$ mV at $T = 293$ K.

This amplifier is used in D/A converter where logarithmic signal is to be reconstructed.

5.6   Logarithmic Multiplier

The logarithmic multiplier utilizes log, summing and antilog amplifiers. The multiplication or division of two signals can be done using the log and antilog amplifiers. Basically we have to multiply two numbers. These two numbers in the form of voltages (V_1 and V_2) are available. Take the logarithm of these two numbers and then add them by the adder. Finally take the antilog of the adder output to get the required multiplication of V_1 and V_2. Following Fig. 5.9 shows how the multiplication of two signals V_1 and V_2 can be obtained.

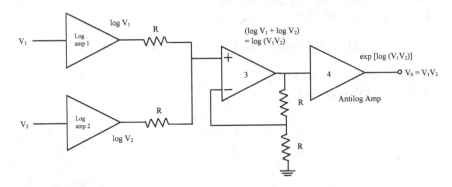

Fig. 5.9 Logarithmic multiplier

Let V_1 and V_2 be the inputs whose product is V_1V_2. In the circuit, Op-Amp 1 and 2 operates as logarithmic amplifier and Op-Amp 3 and 4 as summing or adder and antilog amplifier, respectively.

The outputs of the logarithmic amplifier are given by

$$V_{01} = -nV_T \log\left(\frac{V_1}{RI_s}\right)$$

$$= -B_1 \log\left(\frac{V_1}{B_2}\right) \quad \because B_1 = nV_T \text{ and } B_2 = RI_s \qquad (5.13)$$

Similarly

$$V_{02} = -nV_T \log\left(\frac{V_2}{RI_s}\right)$$

$$= -B_1 \log\left(\frac{V_2}{B_2}\right) \qquad (5.14)$$

Equations (5.13) and (5.14) are the inputs of the adder or summing amplifier. Thus the output of adder is

$$V_0' = -B_1 \left[\log\left(\frac{V_1}{B_2}\right) + \log\left(\frac{V_2}{B_2}\right)\right]$$

$$= -B_1 \log\left(\frac{V_1V_2}{B_2^2}\right) \qquad (5.15)$$

The output of summing amplifier V_0' is the input of antilog or exponential amplifier. Hence

$$V_0 = B_2 \exp\left[\frac{V_0'}{B_1}\right] \qquad (5.16)$$

Substituting the value of V_0' from (5.15) into (5.16), we get

$$V_0 = B_2 \exp\left[-\frac{B_1}{B_1}\log\left(\frac{V_1V_2}{B_2^2}\right)\right]$$

$$V_0 = -\frac{V_1V_2}{B_2} = -\frac{1}{B_2}V_1V_2 \qquad (5.17)$$

Thus the output of the multiplier is $\left(-\frac{1}{B_2}\right)$ times product of two inputs V_1 and V_2. The final output can be obtained by an inverting amplifier having voltage gain $(-B_2)$.

The division of the input signals can be performed by subtracting the outputs of the log amplifier. The multiplier or divider is useful only for unipolar inputs. This is called one quadrant operation.

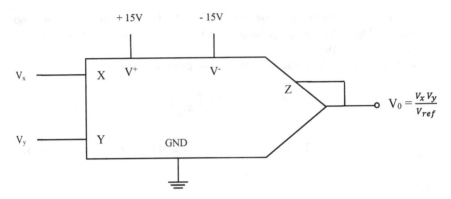

Fig. 5.10 Analog multiplier symbol

By taking the same inputs as $V_1 = V_2 = V$, the squaring of input voltage can be obtained using analog multiplier.

The symbol (Fig. 5.10) for analog multiplier is shown below.

5.7 Amplitude Modulator (AM)

The amplitude modulation can be obtained by analog multiplier. In modulation two signals, namely modulating input and carrier input, are present. It is a process by which the sound waves or signal is transmitted over a carrier wave. This is generally used for communication of sound signal or audio signal over a long distance.

Let $V_1 = V_c \cos \omega_c t$ and $V_2 = V_m \cos \omega_m t$ be the carrier and modulating input signals, respectively, where V_c and V_m are the carrier and modulating signals amplitudes, respectively, and ω_c and ω_m are the angular frequencies of the respective signals.

Then the output of balanced modulator or multiplier can be obtained as

$$V_0 = V_c V_m \cos \omega_c t \, \cos \omega_m t$$

$$= \frac{V_c V_m}{2} [\cos(\omega_c + \omega_m)t + \cos(\omega_c - \omega_m)t] \tag{5.18}$$

The output of multiplier represents the AM modulated signal. Figure 5.11 represents the balanced modulator.

The above Eq. 5.18 has $(\omega_c + \omega_m)$ and $(\omega_c - \omega_m)$ frequencies. The carrier frequency (ω_c) and modulation frequency (ω_m) terms are absent in Eq. 5.18. The two discrete frequencies $(\omega_c + \omega_m)$ and $(\omega_c - \omega_m)$ are newly appeared at the output which are known as upper and lower side bands. The carrier frequency component (ω_c) is completely suppressed at the output signal. Hence, it is called suppressed carrier modulation. It is the basic property of the balanced modulator to suppress any one of the input signals at the output signal.

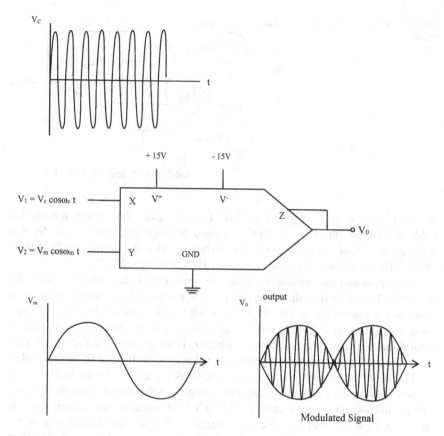

Fig. 5.11 Balanced modulation

5.7.1 Sample and Hold Circuit

In any data acquisition system the data receives from different sources and transmits this data to the next circuit for processing. The circuit in which the data receiving from preceding section samples or captures fromcontinuously varying signal and holds or locks for a specified minimum period till next input signal is sampled again is called sample and hold circuit. This circuit is useful in A/D converters and digital interfacing systems. Mostly in telephone systems or communication or computer this type of circuit is very useful.

Figure 5.12 shows the simple sample and hold circuit using Op-Amp.

Sample and hold circuit, sample an analog signal and stores this signal for some time. This circuit is also called track and hold circuit. Generally it is a switch. Enhancement-type N-channel MOSFET is generally used in the design of circuit as a switch. Because enhancements MOSFET do not have any conducting channel between source and drain region as in depletion MOSFET. The E-MOSFET

Fig. 5.12 Sample and hold circuit

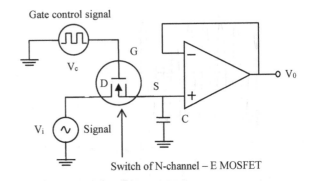

Gate control signal

conducts only when voltage is applied between gate and source region, i.e. E-MOSFETS are off if no voltage is applied between gate and source. So this prevents the false triggering of the switching circuit. Moreover N-channel MOSFETS have faster switching time than P-channel MOSFETS.

The sample and hold circuit is a switch in series with a capacitor C. Initially the E-MOSFET is off by default. When input V_i is applied and logic control signal V_c is given to the gate of the E-MOSFET, the E-MOSFET turns on, and the capacitor tracts the input signal during time T_S and charges it to the instantaneous value of input signal V_i with a time constant $r_d C$ where r_d is the resistance of the E-MOSFET when it is on. During this time the input signal is hold till the end of the interval of T_S. This signal holds up to the next gate control signal arrives, i.e. up to the switch is open. This voltage across the capacitor appears at the output through Op-Amp voltage follower. During this time E-MOSFET is off because gate control signal is zero. The capacitor will not discharge because of high input impedance of the Op-Amp voltage follower. The capacitor holds the input signal across it. The time period during which input signal retained by the capacitor is equal to the magnitude of the input signal is called sample period T_S. The time period during which the signal holds across the capacitor is called hold period T_H.

For the better operation of the circuit the frequency of gate control signal should be significantly higher than the input signal. Generally low leakage resistance capacitors of polystyrene, polyethylene and Teflon dielectric materials are preferred for better results. Following factors influence the operation of the sample and hold circuit.

(1) The dielectric absorption phenomenon in capacitor
(2) Input bias current of the Op-Amp
(3) Aperture time
(4) Acquisition time
(5) Slew rate of the Op-Amp.

Chapter 6
Waveform Generators and Comparators

The circuits for wave generation of sinusoidal, square and triangular are described. The concept of negative feedback and relevant theory in brief is described in this chapter. Classifications of oscillators such as phase shift, Wein Bridge, Colpitts and Hartley oscillator and non-sinusoidal oscillators such as astable and monostable multivibrators, their working, function generator, comparator and Schmitt trigger are reported. More attention is paid to the pedagogy, explanation and working of circuits of these topics. The student-centric approach is given.

6.1 Introduction

The *oscillators* are important part of many electronic instruments and equipments. In many devices some fixed frequency oscillators are necessary to generate the sinusoidal and/or pulsating signals. For example the microprocessor kit requires fixed frequency signal of MHz which can be generated from crystal oscillator or a clock pulse is a square wave signal needed for digital circuits. The signal transmitters where modulating signal is generated and the oscillator having intermediate frequency are needed. Therefore the study of oscillators and the generation of signal of variable frequency are utmost important in electronics. The signal can be generated by various methods, but the circuits constructed using operational amplifiers are discussed in this chapter. The working principles of various types of oscillators and the sinusoidal and the non-sinusoidal waveform generation circuits are discussed with relevant theory.

The circuit which generates a waveform of various frequencies of fixed amplitude without any external signal is called *oscillator*. It generates alternating current or voltage waveforms. The feedback is used for the generation of waveform.

Feedback is a process in which a fraction or the part of the output signal is fed back to the input.

There are two types of feedback in amplifier.

(i) Positive feedback (ii) negative feedback

© The Author(s), under exclusive license to Springer Nature Singapore Pte Ltd. 2022
S. Yawale and S. Yawale, *Operational Amplifier*,
https://doi.org/10.1007/978-981-16-4185-5_6

Fig. 6.1 Block diagram of oscillator

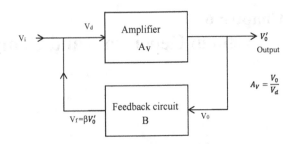

If the input signal and feedback signal are in phase which cause increase in output, it is known as *positive feedback*.

If the input signal and feedback signal are out of phase which cause decrease in output, it is known as *negative feedback*.

The *positive feedback* is used to generate the *oscillations*. This will increase the gain of the circuit. This feedback is called *regenerative feedback*.

In this chapter we are limited to the positive feedback only. Therefore negative feedback is not discussed in this chapter. The negative feedback is discussed in earlier chapter.

Oscillator is a positive feedback amplifier in which part of the output is given as feedback to the input through a feedback network. Following Fig. 6.1 shows the positive feedback amplifier or *oscillator block diagram*.

From block diagram, open loop gain of the amplifier A_V is given by

$$A_V = \frac{V_0}{V_d} \tag{6.1}$$

$$V_0 = A_v V_d$$

where V_0 is the output before feedback.

And the feedback voltage V_f is given by

$$V_f = \beta V_0' \tag{6.2}$$

The input voltage is $(V_i + V_f)$; when this voltage is amplified by A_V times, then it becomes $A_V (V_i + V_f)$.

Therefore after positive feedback the output of the amplifier will be

$$V_0' = A_v(V_i + V_f) \tag{6.3}$$

$$= A_v\left(V_i + \beta V_0'\right)$$

$$\therefore \quad V_0' = A_v V_i + \beta V_0' A_v$$

$$V_0' - \beta V_0' A_v = A_v V_i$$

$$V_0'(1 - \beta A_v) = A_v V_i$$

Or

$$\frac{V_0'}{V_i} = \frac{A_v}{(1 - \beta A_v)} \tag{6.4}$$

where β is called feedback factor or gain and $(1 - \beta A_v)$ is called loop gain.

When $V_i = 0$, then $\beta A_v = 1$.

The positive feedback increases the gain of the amplifier; hence, it is called regenerative feedback. When $\beta A_v = 1$, then mathematically from Eq. (6.4) the gain becomes infinite which means that there is an output without any input. Thus the amplifier becomes an oscillator when $\beta A_v = 1$ and the phase shift of loop gain βA_v must be 0° or 360°.

6.2 Frequency Stability

It is an ability of the circuit to produce exact frequency. The output frequency of the electronic oscillator depends on various factors such as supply voltage and temperature. But the effect on stability can be minimized by using good regulated power supply and thermal shielding. The stability of the oscillator circuit is decided by the figure of merit of the circuit. Higher the figure of merit, better will be the stability of the oscillator. The crystal oscillators are more stable than the RC or LC or any other types of oscillators. The RC oscillators are used for the generation of low or audio frequencies, and LC or tuned oscillators are used for the generation of high-frequency signals. In computers for generation of clock signals the crystal oscillators are employed.

6.3 Classification of Oscillators

There are various types of oscillators. The electronic oscillators are classified into two groups:

(i) sinusoidal oscillators and (ii) non-sinusoidal oscillators.

6.3.1   The Sinusoidal Oscillators

These are the electronic oscillators which produce sine waveforms. These oscillators are divided into

(a) RC phase shift oscillators
(b) Tuned or LC feedback oscillators
(c) Crystal oscillators
(d) Negative resistance oscillators.

6.3.1.1   (a) Phase Shift Oscillator

In *phase shift oscillator* RC networks consist of three sections which introduces 180° phase shift are used in feedback circuit. Each RC section introduces a phase shift of 60°, i.e. total 180° phase shift is introduced with three RC sections. Figure 6.2 shows the Op-Amp phase shift oscillator. In the circuit Op-Amp is just used for amplification and RC circuit as feedback network.

In the circuit Op-Amp is used in inverting mode so it introduces 180° phase shift at the output signal. The each RC network provides 60° phase shift, thus total phase shift introduced by three sections of RC feedback network is 180°. Hence the phase shift at input signal of Op-Amp is 360° or 0°. Therefore positive feedback is created, and by selecting proper value of feedback resistor R_f, oscillations are generated. When the gain of the amplifier and phase are exact, the signal of proper frequency is generated. Hence the oscillation frequency is given by

Fig. 6.2 Op-Amp phase shift oscillator

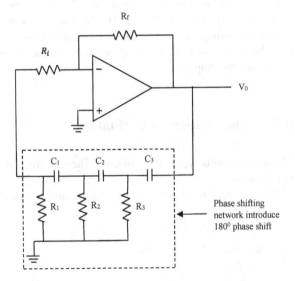

$$f = \frac{1}{2\pi\sqrt{6}RC} \quad \because C = C_1 = C_2 = C_3 \text{ and} \tag{6.5}$$

$$= \frac{0.065}{RC} \quad \because R = R_1 = R_2 = R_3$$

The gain of the amplifier at this frequency must be more than 29.

$$\text{i.e.} \frac{R_f}{R_i} = 29 = \beta \tag{6.6}$$

Thus when gain of the amplifier is adjusted to 29, a desired oscillation of frequency f will be generated.

Advantages of phase shift oscillator.

(i)   The phase shift circuit is simpler than other oscillator circuits.
(ii) The output of this circuit is distortion-free sinusoidal wave.
(iii) This circuit has good frequency stability.
(iv) The value of output sine waveform can be changed by changing the value of R and C simultaneously.

6.3.1.2   (b) Wien Bridge Oscillator

The most popular audio frequency generator is Wien bridge oscillator. This has good frequency stability and low-frequency distortion. The ease of tuning is better; hence, it is often used in laboratory. Figure 6.3 shows the Wien bridge oscillator circuit. It is a RC tuneable oscillator.

The circuit consists of series combination of R_1 and C_1 connected in output and non-inverting input terminal of Op-Amp. The parallel combination of R_2 and C_2 is connected between non-inverting input of Op-Amp and other terminal to the

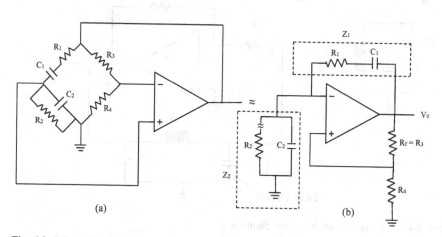

Fig. 6.3 Wien bridge oscillator circuit

ground. The feedback is provided through $R_f = R_3$ and R_4 voltage divider arrangement to the inverting input of Op-Amp. For the ease of convenience, Fig. 6.3a is redrawn and is shown in Fig. 6.3b.

As per the previous discussion the total phase shift must be 0°, and this occurs only when the bridge is balanced. The bridge is balanced at the resonance. When this condition is satisfied, oscillations are generated at the output of the circuit.

The frequency of oscillation is given by

$$f = \frac{1}{2\pi RC} \tag{6.7}$$

The feedback network provides gain of $\frac{1}{3}$,

$$\text{i.e. } A_v = \frac{1}{\beta} = 3 \text{ or } 1 + \frac{R_f}{R_4} = 3 \text{ or } R_f = 2R_4 \tag{6.8}$$

It is evident that when $R_f = 2R_4$, the bridge is balanced at the frequency $f = \frac{1}{2\pi RC}$. In practice there must be a small imbalance of the bridge. But for open loop gain of the amplifier, the bridge is closer to the balance condition at this moment frequency stability of the oscillator is greater.

6.3.1.3   (C) LC Tunable Oscillator (Colpitts Oscillator)

In this oscillator the LC tank circuit is used in feedback network. The tapped capacitance is used as shown in Fig. 6.4a, b.

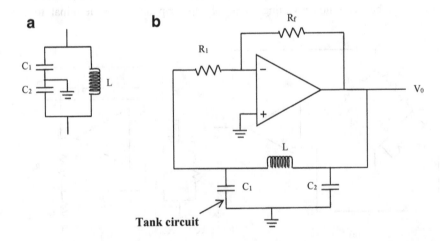

Fig. 6.4 a LC tank circuit, **b** Colpitts oscillator

The Op-Amp works as an inverting amplifier with high gain and LC network as positive feedback. The circuit action of LC tank circuit is same as ordinary LC tank circuit, where the small noise voltage is amplified by Op-Amp when power supply is on. This amplified voltage will charge the capacitor. The signal occurring across the capacitor C_1 is fed to the inverting amplifier which gets amplified by the Op-Amp. At the same time the capacitor gets discharge and again charges generating oscillations. The amplifier introduces phase shift of 180°, and feedback also introduces 180°. Thus total phase shift is 360°. The frequency of oscillations is given by

$$f = \frac{1}{2\pi\sqrt{LC}} \quad \text{where } C = \frac{C_1 C_2}{C_1 + C_2} \tag{6.9}$$

This type of oscillator is generally used to generate a signal up to 1 MHz. The frequency can be varied by ganged capacitor C_1 and C_2. These oscillators are used in microwave applications, surface acoustic wave resonator, mobile communication as well as high-frequency generation.

6.3.1.4   (d) Hartley Oscillator

In Hartley Oscillator instead of tapped capacitor single tapped coil having two parts L_1 and L_2 is used. The capacitor is connected across it. Following Fig. 6.5 shows the circuit of Hartley oscillator using Op-Amp.

When power supply is on, the output voltage appearing starts the charging of the capacitor C. Once the capacitor gets fully charged, it discharges through inductors L_1 and L_2 and again starts charging. This will generate the sine waveform which gets amplified through Op-Amp. The mutual inductance caused in the coil is

Fig. 6.5 Circuit of Hartley oscillator

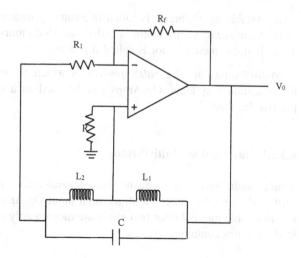

effective which will add to the total inductance of the coils, i.e. $L_1 + L_2 + 2$ M where M is the mutual inductance.

Thus the total inductance will be increased. The frequency of oscillation for this oscillator depends on total inductance and capacitance. In such circuit the feedback network introduces phase shift of 180°, and the reversal occurs due to opposite ends of inductors L_1 and L_2 in combination with the tap connected to the non-inverting terminal or ground through R. The Op-Amp itself introduces a phase shift of 180°, thus total phase shift becomes 360°, which makes the feedback regenerative or positive.

The frequency of oscillator is the resonance frequency of LC tank circuit which is given by

$$f = \frac{1}{2\pi\sqrt{L_T C}} \text{ where } L_T = L_1 + L_2 + 2M \text{ or } L_1 + L_2. \qquad (6.10)$$

To generate the oscillations, the amplifier gain must be greater than the ratio of two inductances $A_v = \frac{L_1}{L_2}$.

If mutual inductance exists between inductors L_1 and L_2, then gain will be $A_v = \frac{(L_1 + M)}{(L_2 + M)}$.

This oscillator cannot be used for low-frequency generation. In this oscillator harmonic contents are very high.

6.3.2 Non-sinusoidal Oscillators

The oscillators whose output is other than sinusoidal wave are called *non-sinusoidal oscillator*. This category includes the square, triangular, saw tooth, etc., waveforms. *Multivibrators* include in this category.

There are three types of multivibrators.

(i)   Astable multivibrator is called free-running multivibrator.
(ii) Monostable multivibrator is called one shot multivibrator.
(iii) Bistable multivibrator is called flip-flops.

Multivibrators are the *pulse generators* which are extensively used in many instrumentation systems. Op-Amps can be used as a multivibrator with proper positive feedback.

6.3.2.1   (a) Astable Multivibrator

Astable multivibrator is a square wave generator known as *free-running multivibrator*. Figure 6.6a shows the differential input Op-Amp free-running multivibrator. In astable multivibrator two states are momentarily stable and circuit switches in these states continuously.

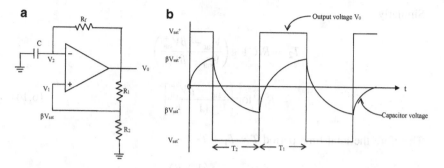

Fig. 6.6 a Symmetrical astable multivibrator, **b** output waveform of symmetrical astable multivibrator

The continuous change in the two states are positive and negative saturation levels. Thus the amplifier output is a square wave. Initially voltage across the capacitor C is zero, i.e. $V_2 = 0$ V. Let the amplifier is in positive saturation so $V_0 = + V_{Sat}{}^+$. Therefore the voltage at points V_2 and V_1

$$V_2 = \frac{R_2}{R_1 + R_2} V_{sat+} \tag{6.11}$$

$$V_1 = \beta V_{sat+} \quad \therefore \beta = \frac{R_2}{R_1 + R_2} \tag{6.12}$$

In astable multivibrator both positive and negative types of feedback are used. The response of negative feedback is slower than the positive feedback, i.e. positive feedback is faster. The voltage $V_1 = \beta V_{sat}{}^+$ which is more positive than the voltage V_2. At this time the capacitor starts charging through resistor R_f.

Thus the voltages V_1 and V_2 get compared and as soon as $V_1 > V_2$, the output of the amplifier suddenly switches to $V_{sat}{}^-$ due to regeneration action. At the same time the capacitor C starts discharging and recharging in opposite direction. Again the Op-Amp compares the voltages V_1 and V_2, the moment V_2 crosses V_1, the state of the multivibrator changes from $V_{sat}{}^-$ to $V_{sat}{}^+$. The charging and discharging action of capacitor is slow; hence, the saturation voltage remains constant. In this way the state of the multivibrator is changing continuously with time, and it will change its state from $V_{sat}{}^+$ to $V_{sat}{}^-$ and so on.

The output waveforms will help us to understand the action of astable multivibrator.

The capacitor charging period T_1 and T_2 can be calculated as

$$T_1 = R_f C \log \left(\frac{V_{sat}^+ - \beta V_{sat}^-}{V_{sat}^+ - \beta V_{sat}^+} \right)$$

$$= R_f C \log \left[\frac{V_{sat}^+ - \beta V_{sat}^-}{V_{sat}^+ (1 - \beta)} \right] \tag{6.13}$$

Similarly

$$T_2 = R_f C \log \left(\frac{V_{sat}^- - \beta V_{sat}^+}{V_{sat}^- - \beta V_{sat}^-} \right)$$

$$= R_f C \log \left[\frac{V_{sat}^- - \beta V_{sat}^+}{V_{sat}^-(1 - \beta)} \right] \tag{6.14}$$

Therefore the total time period $T = T_1 + T_2$

$$T = 2R_f C \log \left(\frac{1 + \beta}{1 - \beta} \right) \tag{6.15}$$

Or

$$f_0 = \frac{1}{2R_f C \log \left[1 + \left(2\frac{R_2}{R_1} \right) \right]} \qquad \therefore \beta = \frac{R_2}{R_1 + R_2} \tag{6.16}$$

Equation (6.16) represents the frequency of output waveform. The frequency can be varied by varying the value of resistor R_f. For getting the output symmetrical two Zener diodes can be connected back to back in output terminal and ground.

In the above multivibrator, when time $T_1 = T_2$, then it is called *symmetrical multivibrator*. When time $T_1 \neq T_2$, then it is called as *asymmetrical multivibrator*. Following Figs. 6.7 and 6.8 shows the circuit diagram and waveform of asymmetrical multivibrator, respectively. The capacitor C charges through diode D_1 and resistor R_3 in forward direction. At this time diode D_2 is reverse biased during T_1.

Fig. 6.7 Asymmetrical astable multivibrator

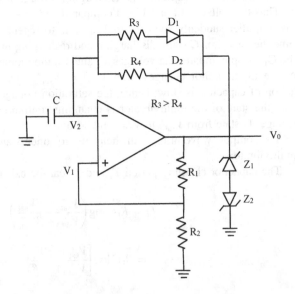

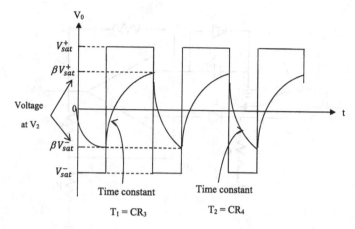

Fig. 6.8 Waveform of asymmetrical astable multivibrator

It will charge in a reverse direction through R_4 and D_2 during T_2. Of course voltage at V_2 will change as per the forward and reverse bias of diodes D_1 and D_2, hence charging of capacitor C in that direction. This constitutes total time $T = T_1 + T_2$.

6.3.2.2 (b) Monostable Multivibrator

Monostable multivibrator called as *one shot* has only one stable state. This stable state can be changed to other states by applying a trigger pulse, but it will come back to its stable state after a specific time T, which decides by R and C values.

Following Fig. 6.9 shows the circuit of Op-Amp monostable multivibrator.

Assume that in a stable state of the circuit the output V_0 is positive saturation, V_{sat}^+. The voltage at point A is positive having voltage βV_0 or βV_{sat}^+ where

$$\beta = \frac{R_1}{R_1 + R_2} V_{sat}^+$$

Now if the voltage at point A is brought down to the zero, then circuit will switche to V_{sat}^- level. This will be accomplished by applying a short spike. The amplifier goes to the quasi-stable state. Then the voltage βV_0 becomes, i.e. voltage at point A becomes negative, i.e. βV_{sat}^-. At this time potential at point B falls through capacitor discharging via R_f and diode D_2, which is reverse biased. Now the potential at point B changes slowly and as soon as it becomes βV_{sat}^-, it will change its state. The waveform along with trigger pulse is shown in Fig. 6.10.

The capacitor gets charged through resistance R_f from zero to βV_{sat}^-. Thus the charging period is given by

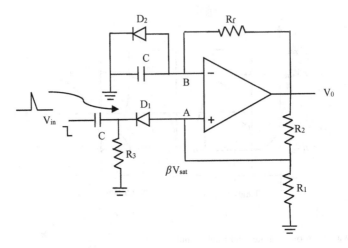

Fig. 6.9 Monostable multivibrator

Fig. 6.10 Waveforms of
monostable multivibrator

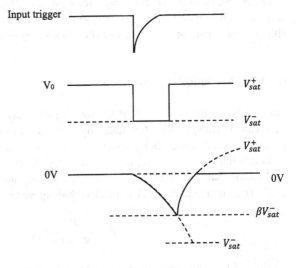

$$T = R_f C \log \left(\frac{V_{sat}^- - 0}{V_{sat}^- - \beta V_{sat}^-} \right) \qquad (6.17)$$

$$T = R_f C \log \left[\frac{V_{sat}^-}{V_{sat}^- (1 - \beta)} \right]$$

$$= R_f C \log \left(\frac{1}{1 - \beta} \right) \qquad (6.18)$$

Substituting the value of β in Eq. (6.18) we get

$$T = R_f C \log\left(1 + \frac{R_1}{R_2}\right) \quad \because \beta = \frac{R_1}{R_1 + R_2} \qquad (6.19)$$

The time period can be set by selecting proper values of capacitor C and resistance R_f.

6.4 Comparators

The Op-Amp when operated under open loop condition, then it would go to the saturation, i.e. either positive $\left(V_{sat}^+\right)$ or negative $\left(V_{sat}^-\right)$ saturation. In such a condition the Op-Amp is said to be operated as a nonlinear device.

Figure 6.11a, b shows the circuit of simple comparator using Op-Amp.

Assuming that the output of the amplifier is $+V_{sat}^+$, it may be equal to $+V_{CC}$ supply voltage. The non-inverting input is supplied by $+1V$, and input V_i is varied from 0 to +ve value greater than $+1V$ (which is at non-inverting input). If the input voltage V_i is gradually increased, then the output remain at V_{sat}^+ till the voltage at point A crosses the voltage $+1V$ at point B, the non-inverting terminal. As soon as it crosses the $+1V$ the output of the comparator suddenly changes from $+V_{sat}^+$ to V_{sat}^-. Further increase in the voltage at point A, the output remains at V_{sat}^- (Fig. 6.11b). The saturation voltage may be equal to $+V_{CC}$ or $-V_{CC}$, the supply voltage.

Such circuit of Op-Amp is called comparator because it compares the voltage with either input terminal of Op-Amp. For ideal comparator the switching should be short as well as input offset voltage should be low.

If the input at non-inverting terminal, i.e. at point B is zero or it is grounded, then voltage applied at inverting terminal will be compared with the zero volts. In this

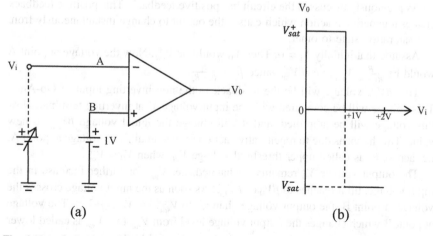

(a)   (b)

Fig. 6.11 a, b Op-Amp comparator

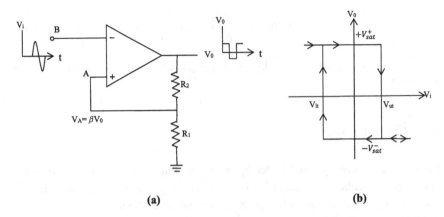

Fig. 6.12 a, b Schmitt trigger and output waveform

case as soon as the input at inverting terminal crosses the zero volt, the output of the comparator suddenly changes its earlier saturation level. Hence the circuit is called *zero crossing detector*. *Zero crossing detector* is used for conversion of sine wave into square waveform.

6.4.1 Schmitt Trigger

Schmitt trigger is an inverting comparator which converts sine wave or any irregular wave into square wave. It is a bistable circuit possesses two stable states. These two stable states are continuously changing as per the proper input supplied at inverting terminal of Op-Amp, due to regenerative feedback.

Figure 6.12a, b shows the circuit of Schmitt trigger using Op-Amp.

As previously discussed the circuit has positive feedback. This positive feedback creates regenerative action which causes the output to change instantaneously from one saturation state to other.

Assume that initially $V_i = 0$. Then V_0 would be V_{sat}^+. Now the voltage at point A would be $\frac{R_1}{R_1 + R_2} V_{sat}^+$, i.e. βV_{sat}^+ since $\beta = \frac{R_1}{R_1 + R_2}$.

This βV_{sat}^+ voltage will be the feedback to the non-inverting input of Op-Amp. This voltage will be compared with the input voltage V_i at inverting terminal. Now this voltage will be amplified, and it will change the initial voltage βV_{sat}^+ to new value. This happens due to regenerative action of the circuit. The voltage at point A, i.e. across R_1 is called upper threshold voltage V_{ut}, when $V_0 = V_{sat}^+$.

The output voltage V_0 remains unchanged, i.e. V_{sat}^+ for further increase in the input voltage, till it reaches to βV_0 or βV_{sat}^+. As soon as the input voltage crosses the voltage at point B, the output voltage changes to V_{sat}^-, i.e. $V_0 = -V_{sat}^-$. The voltage at point B which changes the output voltage level from V_{sat}^+ to $-V_{sat}^-$ is called lower threshold voltage which will be $\frac{R_1}{R_1 + R_2} \left(-V_{sat}^-\right)$ denoted by V_{lt}.

The difference between upper threshold voltage and lower threshold voltage is called hysteresis.

$$V_H = V_{ut} - V_{lt}$$

$$= \beta V_{sat}^+ \left(-\beta V_{sat}^- \right)$$

$$= \beta \left(V_{sat}^+ - V_{sat}^- \right)$$

$$= \frac{R_1}{R_1 + R_2} \left(V_{sat}^+ - V_{sat}^- \right) \tag{6.20}$$

The speed of operation and accuracy are the main important characteristics of comparator. The speed of operation depends on the bandwidth of the Op-Amp. Higher the bandwidth, more will be the speed of operation. Similarly accuracy depends on the input offset voltage and current, thermal drifts and CMRR. Smaller the hysteresis, the better would be the response of the comparator.

6.5 Waveform Generators

The *astable multivibrator* generates square wave, and integrator circuit converts this square wave to the *triangular wave*. Simply the output of astable or free-running multivibrator is given to the input of integrator produces triangular waveform. If on time and off time of square wave are equal, the triangular wave of equal ramp up and down signal is generated. If on time is smaller than the off time of the square wave, i.e. $T_{OFF} > T_{ON}$, the sawtooth waveform is generated.

Another method of triangular wave generator is very popular in which zero crossing detector is used as comparator for square wave generation (Fig. 6.13a, b).

Obviously two Op-Amps are required one for comparator and other for integrator. The comparator always compares the voltage at non-inverting input with 0 V at inverting input because inverting input of Op-Amp is grounded.

The voltage appearing at point N is compared with the ground potential, thus the output of Op-Amp A, i.e. comparator changes from $+V_{sat}$ to $-V_{sat}$ saturation level ($+ V_{sat} = + V_{CC}$ and $-V_{sat} = -V_{CC}$). This square wave signal converts into ramp up and down signal by the integrator (Op-Amp B). Capacitor C is connected in feedback of Op-Amp B.

The amplitude and frequency of triangular wave can be calculated from the various component values in the circuit. From the waveforms in Fig. 6.13 (b), it can be seen that initially the output of the comparator A is positive $V_{sat} = V_{CC}$, and the output of the integrator circuit B decreases and reaches to some value $-V_{TR}$. During this moment the Op-Amp A compares the two voltages and suddenly changes its state from $+ V_{sat}$ to $-V_{sat}$ ($-V_{EE}$). Before change in the state of Op-Amp A, the

a

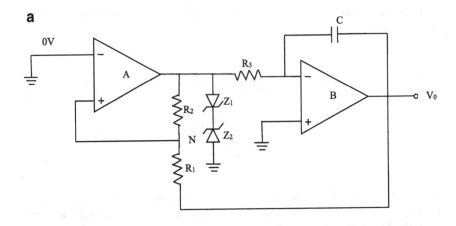

b

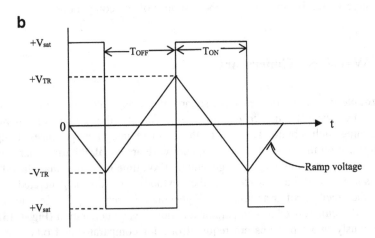

Fig. 6.13 a Triangular wave generator, **b** waveforms at subsequent outputs

voltage at point N was 0 V. This means that the negative ramp voltage ($-V_{TR}$) develops across resistance R_2 and $+V_{sat}$ voltage develops across resistance R_1.

Thus

$$\frac{-V_{TR}}{R_2} = -\frac{+V_{sat}}{R_1}$$

Or

$$-V_{TR} = -\frac{R_2}{R_1}(+V_{sat}) \qquad (6.21)$$

Similarly $+V_{TR}$ is developed across R_2 in Op-Amp A and $-V_{sat}$ across resistance R_1 in Op-Amp B.

$$+V_{TR} = -\frac{R_2}{R_1}(-V_{sat}) \tag{6.22}$$

From Eqs. (6.21) and (6.22), the peak to peak amplitude of triangular waveform is

$$V_{OPP} = +V_{TR} - (-V_{TR})$$

$$= 2\frac{R_2}{R_1}V_{sat} \quad \because V_{sat} = |+V_{sat}| = |-V_{sat}| \tag{6.23}$$

From waveforms it can be seen that the time required for changing output voltage of triangular wave from $-V_{TR}$ to $+V_{TR}$ is T/2. The saturation voltage is the input of integrator, therefore using integrator equation the output frequency can be calculated.

$$V_{OPP} = \frac{1}{R_3C_1}\int_0^{T/2}(-V_{sat})dt$$

$$= \frac{V_{sat}}{R_3C_1}\left(\frac{T}{2}\right)$$

Hence

$$T = 2R_3C_1\frac{V_{OPP}}{V_{sat}} \tag{6.24}$$

where $C = C_1$ and $V_i = -V_{sat}$.
Substituting value of V_{OPP} from Eq. (6.23) into Eq. (6.24), we get

$$T = \left(\frac{2R_3C_1}{V_{sat}}\right)\left(2\frac{R_2}{R_1}V_{sat}\right)$$

$$T = \frac{4R_2R_3C_1}{R_1} \tag{6.25}$$

The frequency of oscillation of triangular waveform is

$$f = \frac{1}{T} = \frac{R_1}{4R_2R_3C_1} \tag{6.26}$$

Above equation shows that the frequency of oscillation depends on resistor R_1. For obtaining the desired output voltage, two Zener diodes are connected back to back at the output of Op-Amp A. The square wave can be obtained at the output of Op-Amp A and triangular wave at the Op-Amp B.

When duty cycle of the astable multivibrator output is same, the triangular waveform will be obtained at the output of integrator.

When duty cycle is less than 50%, i.e. unequal, T_{ON} and T_{OFF} times of the square wave generates saw tooth wave. A small modification in triangular wave generator can be made to change the duty cycle which generates saw tooth waveform. In Op-Amp B, i.e. integrator circuit, instead of grounding the non-inverting input a potentiometer of 20 k ohm whose variable terminal is connected to non-inverting input and extreme ends to the $+V_{CC}$ and $-V_{CC}$ (i.e. power supply). By adjusting duty cycle the rise and fall times of the triangular wave adjusted to get saw tooth wave.

Chapter 7
Active Filter Circuits and Phase-Locked Loop (PLL)

Active filters and phase-locked loop (PLL) and its applications are discussed in this chapter. To get acquainted with the design of active filters and the applicability in instrumentation, low pass, high pass and band pass filters are explained. The frequency selection in audio or music systems is utmost important. The selection of lower and higher cut-off frequencies decides the quality of the filters. And in public address or audio or stereophonic system bass and treble decides the quality of the sound and this comes from the proper selection of low, mid and high frequencies. That is why the mixer circuits are provided in sound recording or music system. Nowadays the equalizers are very common in the television audio and stereophonic system. Thus the frequency selective circuits are extreme important. Classification of active filters and the working of Butterworth first- and second-order filters are described. Phase-locked loop, voltage-controlled oscillator and their operating principle has been used in many applications such as frequency shift keying, decoders, frequency multiplier and translators. The application of PLL as frequency multiplier and translator is described in a lucid language in this chapter.

7.1 Introduction

Active filters play dominant role in many electronic devices. To suppress or to pass a particular frequency or a band of frequencies the active filters are used in many devices. In active filters roll-off frequency is very important one as far as application is concerned. There are various kinds of filters such as Chebyshev, Bessel, Butterworth and elliptical filters. Their characteristics are different as compared to the roll-off and cut-off frequency. These filters are free from insertion loss. Also they provide the isolation between two circuits due to special characteristics such as high input impedance and low output impedance. Filters are applicable in communication electronics, audio amplifier systems, biomedical instrumentation, audio speakers, equalizers, etc. The most important advantage of active filters is to reduce

© The Author(s), under exclusive license to Springer Nature Singapore Pte Ltd. 2022
S. Yawale and S. Yawale, *Operational Amplifier*,
https://doi.org/10.1007/978-981-16-4185-5_7

the complex design of the filter without inductors. Nowadays in stereophonic Dolby audio systems the active filters are playing prominent role.

A frequency selective electric circuit is called filter that attenuates or passes selective band of frequencies. Filters may be classified as follows

(i) Digital filters, (ii) Analog filters, (iii) Passive filters, (iv) Active filters, (v) Audio filters, (AF) (vi) Radio frequency filters, (RF) (vii) video frequency filters, (viii) microwave filters and (ix) ultrahigh frequency filters (UHF).

The analog filters process analog signals, whereas digital filters process digital signal. In a filter circuit if passive components like capacitors and resistors are used for design of the filters, then they are called *passive filters,* and if active components like transistors, Op-Amps along with resistors and capacitors are used then the filters are called *active filters.* As per their working and frequency range the filters are classified. If these are design for audio frequency, then they are called *audio filters* and so on.

7.2   Classification of Active Filters

The active filters are classified into four types as

1. Butterworth filter
2. Chebyshev filter
3. Bessel filter and
4. Elliptical filter.

The basic difference between these types of filters is the change in roll-off. The *roll-off* is nothing but the response of the filter at cut-off frequency. This roll-off of the filter changes with order of filter as first, second, third and so on. More roll-off means approaching the filter response nearer to the ideal one.

The Bessel and Butterworth filters have same behaviour as pass band only. Butterworth filter has more roll-off than Bessel filter. The frequency response of the above classical classified filters is shown in Fig. 7.1 from which the detailed idea about the classified filters can be obtained. The cut-off frequency (f_H) is also different for different types of filters.

The Chebyshev filters have sharper cut-off than Bessel and Butterworth filters, but it has ripple in the pass band. So many times it is also called as ripple filter. These filters are not suitable for audio systems because of nonlinear phase response.

Butterworth filters give flat response in pass band and ample roll-off. Maximally flat in pass band and sharper roll-off are more perfect filters applicable in audio systems or many network circuits.

The Bessel filters has quite linear phase response up to cut-off frequency.

Chebyshev filters gives ripple in the pass band and it severe roll-off at the cost of ripple in the stop band.

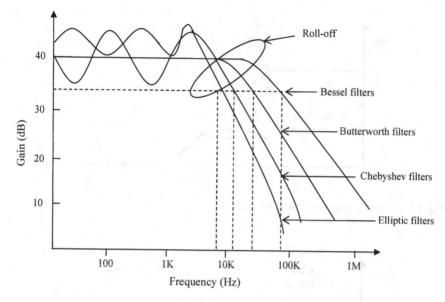

Fig. 7.1 Frequency response of the Bessel, Butterworth, Chebyshev and elliptic filters

The commonly used filter circuits are

1. Low pass filters
2. High pass filters
3. Band pass filters
4. Band reject filters
5. All pass filters

Each of these filters uses Op-Amp as the active element along with resistors and capacitors as passive elements.

1 Low pass filters

Low pass filter is an electronic circuit that passes low frequencies below cut-off frequency and attenuates higher frequencies above cut-off frequency (f_H). Figure 7.2 shows the response of ideal low pass filter. The bandwidth of this filter is (f_H-0) = f_H, the higher cut-off frequency at which the gain of the amplifier is depressed by 3 dB. The gain of the amplifier decreases by attenuating the frequencies greater than f_H.

2 High pass filters

In *high pass filter* the lower frequencies below f_L, the lower cut-off frequency get attenuated and higher frequencies above f_L will be passed or transmitted. Figure 7.3 shows the frequency response of ideal high pass filter. The band of frequencies above f_L are called *pass band*, and the band of frequencies below f_L is called *stop band*. The bandwidth is (f_L-0) = f_L.

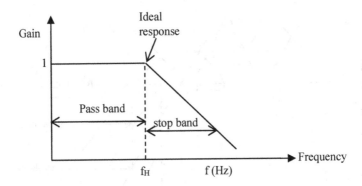

Fig. 7.2 Frequency response of ideal low pass filter

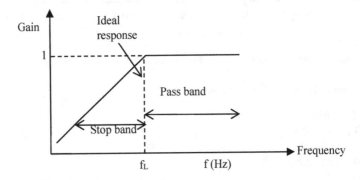

Fig. 7.3 Frequency response of ideal high pass filter

3 Band pass filter

It is an electronic circuit in which the particular band of frequencies will be transmitted and rest of the other frequency band is attenuated. The pass band frequencies are between higher and lower cut-off frequencies i.e. f_H and f_L. Hence the bandwidth is $(f_H - f_L)$. Figure 7.4 shows the frequency response of ideal band pass filter. The higher and lower frequencies decide the pass band limit of band pass filter.

4 Band reject filter

Band reject filter performs exactly opposite function of band pass filter. This device rejects or attenuates the particular band of frequencies and passes rest of the frequencies outside the reject band. In this filter the lower and higher cut-off frequencies will decide the attenuation band. So bandwidth is $(f_H - f_L)$. This filter is

Fig. 7.4 Frequency response
of ideal band pass filter

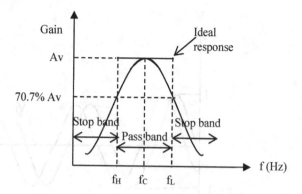

Fig. 7.5 Frequency response
of ideal band stop or notch
filter

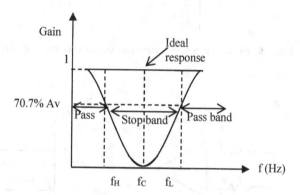

also called *band eliminator or notch or band stop filter*. Figure 7.5 shows the frequency response of ideal band reject filter.

5   All pass filter

In *all pass filters*, all the input signal frequencies will be transmitted at the output with same magnitude but with change in phase. Figure 7.6 shows the phase shift between input and output voltages in all pass filters. This filter is also called *signal processing filter or delay equalizer or phase corrector*.

The gain variation of the filter circuit in the stop band is determined by the order of filter. For example, for the first-order low pass filter, the gain rolls off at the rate of 20 dB/decade in the stop band, for $f > f_H$ and for the second-order low pass filter the roll-off rate is 40 dB/decade, for third order it is 60 dB/decade and so on.

7.2.1   First-Order Low Pass Butterworth Filter

Following Fig. 7.7a shows the first-order low pass Butterworth filter.

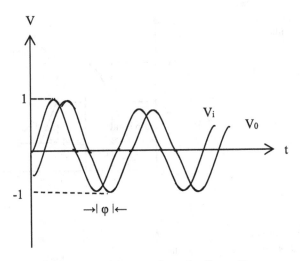

Fig. 7.6 Phase shift between input and output voltages in all pass filters

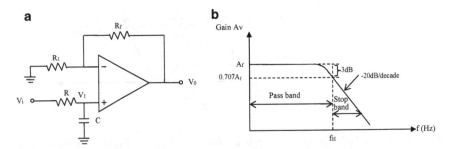

Fig. 7.7 **a** First-order low pass Butterworth filter. **b** Frequency response of first-order low pass Butterworth filter

In Fig. 7.7 a simple RC network is used for filtering. Resistors R_1 and R_f determine the gain of the filter. According to voltage divider rule the voltage at the non-inverting input terminal is

$$V_1 = \frac{-jX_C}{R - jX_C} V_i \qquad (7.1)$$

where $j = \sqrt{-1}$ and $-jX_C = \frac{1}{j\omega C}$.

Simplifying Eq. (7.1), we get

$$V_1 = \frac{V_i}{1 + j\omega RC}$$

and the output voltage

$$V_0 = \left(1 + \frac{R_f}{R_1}\right) \frac{V_i}{1 + j\omega RC}$$

or

$$\frac{V_0}{V_i} = \frac{\left(1 + \frac{R_f}{R_1}\right)}{1 + j\omega RC}$$

$$\frac{V_0}{V_i} = \frac{A_f}{1 + j2\pi f RC} \qquad \because \omega = 2\pi f$$

or

$$\frac{V_0}{V_i} = \frac{A_f}{1 + j\left(\frac{f}{f_H}\right)} \tag{7.2}$$

where $\frac{V_0}{V_i}$ = gain of the filter.

$A_f = 1 + \frac{R_f}{R_1}$ pass band gain of the filter.

f is the frequency of the input signal and $f_H = \frac{1}{2\pi RC}$ is the high cut-off frequency of the filter.

The gain magnitude and phase angle equation of the low pass filter can be obtained by converting Eq. (7.2) into its polar form

$$\left|\frac{V_0}{V_i}\right| = \frac{A_f}{\sqrt{1 + \left(\frac{f}{f_H}\right)^2}} \tag{7.3}$$

$$\phi = -\tan^{-1}\left(\frac{f}{f_H}\right) \tag{7.4}$$

where ϕ is the phase angle in degrees. From Eq. (7.3) three cases arises as $f < f_H$, $f = f_H$ and $f > f_H$. So for $f < f_H$, i.e. when input signal has very low frequency range, in that case Eq. (7.3) gives

$$\left|\frac{V_0}{V_i}\right| \cong A_f \tag{7.5}$$

In the second case when $f = f_H$, the Eq. (7.3) gives,

$$\left|\frac{V_0}{V_i}\right| = \frac{A_f}{\sqrt{2}} = 0.707 A_f = 70.7\% A_f \tag{7.6}$$

And when the input signal frequency $f > f_H$, then the Eq. (7.3) gives

$$\left|\frac{V_0}{V_i}\right| < A_f \tag{7.7}$$

Hence the low pass filter has a constant gain A_f over 0 Hz to f_H, the higher cut-off frequency; at f_H the gain is $0.707 A_f$. Beyond f_H it gradually decreases at constant rate (dB/decade) with an increase in frequency and becoming lesser and lesser as you go on increasing the frequency. The band of frequencies beyond break frequency (Fig. 7.7b) is called stop band. The roll-off is -20 dB/decade.

When the operating frequency is equal to cut-off frequency and beyond the frequency is increased one decade, the voltage gain of the circuit is divided by 10 means it decreases by 20 dB/decade or 6 dB/octave each time. This is called the gain rolls off after higher cut-off frequency f_H (frequency $f = f_H$ at which voltage gain down by 30 dB) is 20 dB/decade or 6 dB/octave. So the pass band gain in dB will be A_v (dB) = 20 log (V_o/V_i).

The cut-off frequency is also called as -3 dB frequency, break frequency or corner frequency.

The conversion of original cut-off frequency to a new cut-off frequency is called frequency scaling. This can be done by changing the value of R and C. But to change the value of R is easier than C, because the proper or exact value of capacitor will not be available in the market. The cut-off or break frequency can be changed by taking the ratio of original cut-off frequency to new cut-off frequency and multiply this ratio with the resistor value R gives you the new value of R to be connected in the circuit. This will give you the new cut-off frequency or -3 dB or break frequency.

7.2.2   Second-Order Low Pass Butterworth Filter

When we obtain the roll-off of 40 dB/decade in a first-order low pass Butterworth filter or cascading the two stages of first-order filter, then it becomes second-order low pass filter. It is a two-pole low pass filter. By cascading various orders of active filters result in to n^{th} order filter. Following Fig. 7.8a shows the circuit diagram of second-order low pass Butterworth filter. The modification that would be made in the first-order low pass filter is to connect the additional RC network as in Fig. 7.8a. The higher cut-off frequency can be determined from the values of R_1, C_1, R_2 and C_2 which is given by

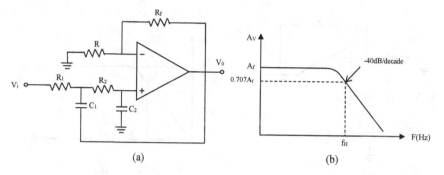

Fig. 7.8 **a** Second-order low pass Butterworth filter and **b** frequency response

$$f_H = \frac{1}{2\pi\sqrt{R_1 R_2 C_1 C_2}} \tag{7.8}$$

whereas the gain remain as it is in first-order circuit which can be adjusted by R_1 and R_f.

The gain of the circuit is given by

$$\left|\frac{V_0}{V_i}\right| = \frac{A_f}{\sqrt{1 + (f/f_H)^4}} \qquad \because A_f = 1 + \frac{R_f}{R_1} \tag{7.9}$$

7.2.3 Fist-Order High Pass Butterworth Filter

By interchanging the positions of capacitor and resistor in first-order low pass filter, high pass filters can be designed. Following Fig. 7.9a, b shows the first-order high pass Butterworth filter and its frequency response.

The R and C values determined the lower cut-off frequency f_L. At this frequency the gain of the amplifier falls to 70.7% of its maximum gain A_f. In this filter the frequencies higher than lower cut-off frequency f_L are passed. The value of f_L can be calculated from $f_L = \frac{1}{2\pi RC}$, where f is the frequency of the signal.

The gain of the circuit is given by

$$\left|\frac{V_0}{V_i}\right| = \frac{A_f(f/f_L)}{\sqrt{1 + (f/f_L)^2}} \qquad \because A_f = 1 + \frac{R_f}{R_1} \tag{7.10}$$

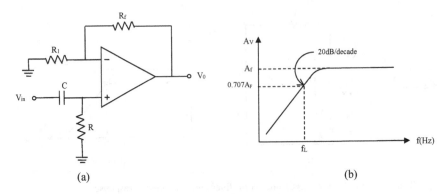

Fig. 7.9 a First-order high pass filter and **b** frequency response

7.2.4 Second-Order High Pass Butterworth Filter

In second-order high pass Butterworth filter the additional RC network is used in first-order circuit. Following Fig. 7.10a, b shows the second-order high pass Butterworth filter along with frequency response.

The roll-off in the second-order high pass Butterworth filter is 40 dB/decade. The frequency f_L is given by

$$f_L = \frac{1}{2\pi\sqrt{R_1 C_1 R_2 C_2}} \quad \text{and gain is} \quad \left|\frac{V_0}{V_i}\right| = \frac{A_f}{\sqrt{1 + (f_L/f)^4}} \tag{7.11}$$

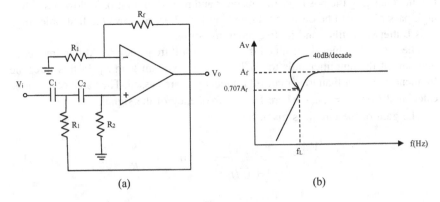

Fig. 7.10 a Secondorder high pass filter and **b** frequency response

7.2.5 Band Pass Filter

In band pass filter the frequencies between lower cut-off and higher cut-off are passed and rests of the others are attenuated. This filter can be design by cascading the high pass and low pass filters. Following Fig. 7.11a, b shows the band pass filter using Op-Amp and its frequency response.

The lower and higher cut-off frequencies in high and low pass filters are set as per the following formulae.

$$f_H = f_L = \frac{1}{2\pi RC} \text{ in first-order filter.}$$

The band width of the filter is $(f_H - f_L)$.

In band pass filter the figure of the merit is important. This will decide the quality of the filter. The figure of merit is given by

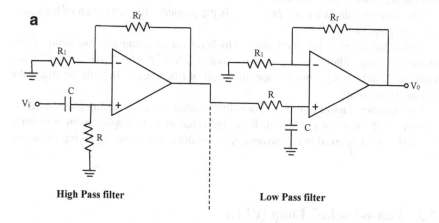

High Pass filter   **Low Pass filter**

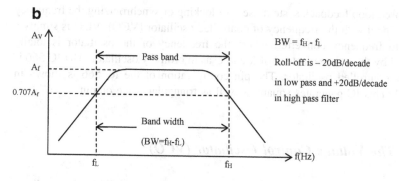

BW = f_H - f_L

Roll-off is – 20dB/decade

in low pass and +20dB/decade
in high pass filter

Fig. 7.11 a Band pass filter. **b** Frequency response

$$Q = \frac{\text{Centre frequency}}{\text{Band width}} = \frac{f_c}{BW} \tag{7.12}$$

or

$$Q = \frac{f_c}{f_H - f_L} \tag{7.13}$$

The centre frequency is given by $f_c = \sqrt{f_H f_L}$, where f_H and f_L are the higher and lower cut-off frequencies, respectively.

If the quality factor or figure of merit Q is greater than 10, i.e. $Q > 10$, the filter circuit is called narrow band pass filter, and if the Q is less than 10, i.e. $Q < 10$, the filter is called wide band pass filter.

If two filters namely first-order high pass and first-order low pass filters are cascaded the roll-off is ± 20 dB/decade, and if second-order filters are cascaded the roll-off becomes ± 40 dB/decade.

The voltage gain in a band pass filter is the product of voltage gain of high pass and low pass filters.

The other filters are a *notch filter* which rejects or eliminates the band of frequencies. These filters are called *band reject filters. Bridge-T and twin-T* notch filters are used for rejecting the specific band of frequencies. In both the filters the roll-off is adjustable.

One another class of filter is all pass filters called phase-correcting filters used for correction of phase of the signal. It is applicable in time displacement functions. Generally used in speaker crossovers, reverberators, telephone signal transmission, etc.

7.3 Phase-Locked Loop (PLL)

It is a closed loop feedback system used for locking or synchronizing the frequency of input signal with the frequency of controlled oscillator (VCO). VCO is simply a voltage to frequency converter, where the frequency of the oscillator is totally controlled by the voltage instead of R and C as in ordinary oscillator. The IC 566 is a voltage-controlled oscillator. The pin configuration of the IC 566 is shown in Fig. 7.12. This IC generates square as well as triangular wave output.

7.3.1 The Voltage Control Oscillator (VCO)

The VCO consists of (i) current source, (ii) Schmitt trigger and (iii) buffer amplifier.

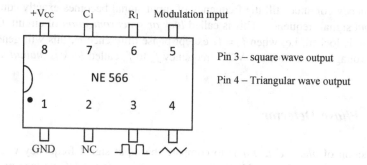

Fig. 7.12 Pin configuration of VCO NE 566

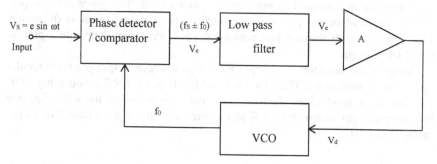

Fig. 7.13 Block diagram of PLL

The block diagram of *phase-locked loop (PLL)* circuit is shown in Fig. 7.13. It consists of phase detector, low pass filter, an error amplifier and voltage-controlled oscillator (VCO).

As previously discussed the frequency of VCO is controlled by external components R_1 and C_1. It is a free-running multivibrator which generates a set frequency f_0 called as free-running frequency. This is the centre frequency which can be shifted to either side by applying DC control voltage V_C. The frequency deviation is directly controlled by the V_C. Therefore it is called *voltage control oscillator* (VCO).

An input signal of $V_S = e \sin \omega t$ of some frequency is applied to the phase detector first which will compare both the phase and frequency of input signal with the output signal coming from VCO. If the signal differs in phase and frequency, an error voltage V_e generated. This signal has frequency components $(f_S \pm f_0)$. The frequency components $(f_S + f_0)$ attenuated by the low pass filter and the $(f_S - f_0)$ components passed through low pass filter get amplified by error amplifier A. The *error signal* (V_d) generated by amplifier is fed to the VCO. This signal V_d shifts the VCO frequency in such a manner to reduce the frequency difference between f_S and f_0, i.e. $(f_S - f_0)$. This action of PLL is called *capture range*. The action of changing the frequency continues till the frequency of input signal becomes exactly equal to the output signal frequency. This is called *lock or synchronization* of circuit. Once the signal is locked, i.e. when $f_0 = f_S$ except phase difference ϕ, this will generate control voltage V_d to shift the VCO frequency f_0 to f_S called *lock is maintained.*

7.3.2   Phase Detector

The function of *phase detector* is to compare the input signal frequency with the voltage-controlled oscillator (VCO) frequency to generate the DC voltage proportional to the phase difference between two frequencies. There are two types of phase detector (i) analog and (ii) digital.

Analog phase detector is a one type of mixer circuit. The phase detector produces a series of output pulses whose width is proportional to phase difference. These pulses passes through low pass filter which smoothen them into a proportional DC voltage.

Digital phase detectors are simply Ex-OR gate whose output produces regular square wave oscillations. The edge-triggered flip-flop, i.e. RS flip-flop using NOR gate can be a good phase detector. The main advantage of the edge-triggered flip-flop phase detector over Ex-OR phase detector is that it gives linear DC output voltage over 360°.

7.3.3   Low Pass Filter

The Op-Amp low pass active filter is generally preferred over conventional RC filter because an amplifier amplifies the signal with required gain. Also it controls the dynamic characteristics of the PLL.

7.4   Applications of PLL

Phase-locked loops are used in many applications such as clock generator, FM demodulation, frequency division and multiplication, frequency translators, ultrafast frequency synthesizers in vector network analysers, motor speed controls, angle modulation, phase demodulator, AM demodulators and clock recovery circuits. Out of these applications frequency multiplier and translators are discussed.

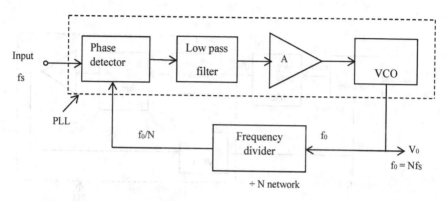

Fig. 7.14 Frequency multiplication using PLL

7.4.1 *Frequency Multiplier*

The PLL can be used as *frequency multiplier*. Figure 7.14 shows the block diagram of frequency multiplier.

A divide by N-network is connected between VCO output and phase detector. The divide by N-network gives output f_0/N, when VCO is in locked state. So input to the frequency divider is lock frequency f_0. Therefore the output frequency $f_0 = Nf_s$.

7.4.2 *Frequency Translation*

It is known that the output of the mixer or comparator contains sum and difference of f_s and f_0, i.e. input frequency and capture frequency. The output of low pass filter is only difference, i.e. $(f_0 - f_s)$. The block diagram of frequency translator is shown in Fig. 7.15.

The translator frequency signal f_1 is used as a second input to the phase detector in PLL. When the PLL is in locked state, the two signals applied to the phase detector are held at same frequency.

Thus

$$f_0 - f_s = f_1$$

or

$$f_0 = f_s + f_1$$

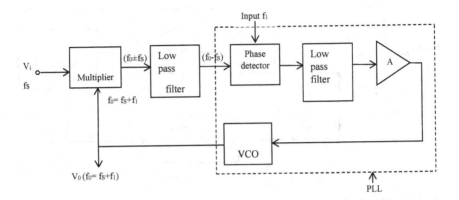

Fig. 7.15 Frequency translator

This means that the VCO signal frequency is translated by f_1.

PLL can be used for amplitude modulation (AM), detection purpose or as demodulator. Also it can be used in FM demodulation (FSK), frequency shift keying (FSK) demodulator.

Chapter 8
Frequency-Dependent Negative Resistance and Gyrator

This chapter contains frequency-dependent negative resistance and gyrator. The gyrator is an important element in integrated circuits performs the work of inductor. The real inductors cannot be fabricated in the ICs, but this could be possible with gyrators. Gyrator is a simulated circuit of inductor. Large value inductors and extensive adjustable inductance range are easily implemented using resistors, capacitors and operational amplifiers; such a circuit is called gyrator. It is a two-port device or network element called hypothetical fifth linear lossless passive element after inductor, capacitor, register and ideal transformer. The theory of gyrator and its working is described in this chapter. Replacement of real iron and air-core inductors by gyrator is discussed. The advantages and disadvantages are also reported. During those days it was an unbelievable thing that inductor without turns of wire on any core. Similarly frequency-dependent negative resistance (FDNR) or D-element is an active element which exhibits real negative resistance views like unusual capacitor. This can be fabricated using resistors, capacitors and operational amplifiers. LC low pass and high pass filters can be transformed using FDNR. In a transformation capacitors are replaced by FDNR inductance by resistance and resistance by capacitor. Gyrator is called synthetic inductor also. It is a negative impedance inverter whose input impedance is proportional to negative of the load admittance.

8.1 Introduction

Inductor is a passive component which cannot be fabricated in integrated circuits (IC). There are few methods of fabrication, but it has some disadvantages like picking up of electromagnetic waves radiating energy and then acts like antenna. This may results in undesirable noise. Basically inductor is designed by core which consists of magnetic material and a conductor windings. This can results in very poor quality factor while fabricating on surface of the semiconductor IC. The

S. Yawale and S. Yawale, *Operational Amplifier*,
https://doi.org/10.1007/978-981-16-4185-5_8

Fig. 8.1 Electrical symbol of gyrator

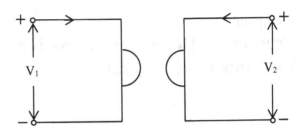

physical size and quality factor of the inductor are the main problems. Harmonic distortion and radiation of electromagnetic waves are main cause which disturbs the working of entire circuit while it is fabricated on surface of the IC using non-planer technology. But it is very difficult to implement a practical inductor with large inductance value and small physical size. Also its value cannot be adjusted exactly. The simulation technique can resolve the problem of fabrication of practical inductor in integrated circuits. This gives the better performance as good as practical inductor. The large inductance value and extensive adjustable inductance range are easily implemented by using resistors, capacitors and operational amplifier called *gyrator*.

The gyrator was first proposed by Bernard D. H. Tellegen in 1948. It is two-port network element. It is also called hypothetical fifth linear lossless passive element after inductor, capacitor, resistor and ideal transformer. The electrical symbol proposed by Tellegen for gyrator is shown in Fig. 8.1.

8.2 Theory of Gyrator

Theory of gyrator and realization of network elements such as capacitor, resistor and inductor is discussed by B.D.H. Tellegen in 1948. He has derived the complex equations of four pole network between the voltages and the currents of the terminals as

$$V_1 = Z_{11}I_1 + Z_{12}I_2 \tag{8.1}$$

$$V_2 = Z_{21}I_1 + Z_{22}I_2 \tag{8.2}$$

Figure 8.2 shows the four-pole network.

Fig. 8.2 Four-pole network

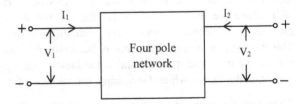

Where four pole parameters Z_{11}, Z_{12}, Z_{21} and Z_{22} are the fractions of frequency.

B. D. H. Tellegen was basically electrical engineer and inventor of gyrator and pentode in the vacuum tube. He is known for his *Tellegen theorem* in the circuit theory of networks. He was working at Phillips Physics Laboratory in Eindhoven in 1926 and was adjunct professor of circuit theory at the University of Delft from 1946 to 1966. He has said that, while deriving the *four-pole network* equations that these network elements have their origin in physics rather than engineering.

The four-pole network violating reciprocity relations is described by

$$V_1 = -si_2$$

$$V_2 = si_1 \qquad\qquad (8.3)$$

In fact from Eq. (8.3),

$$i_1 V_1 + i_2 V_2 = 0 \text{ should be followed.}$$

The coefficients in Eq. (8.3) are not equal as required for reciprocity relations, but they are oppositely equal. Such a four-pole network is denoted by the name of an ideal gyrator.

This shall be considered as a fifth network element. The ideal gyrator gyrates a current into a voltage and vice-versa. The coefficient 's' is called *gyration resistance*, and '1/s' is called *gyration conductance*.

In Fig. 8.2 if secondary terminal is open, i.e. $i_2 = 0$, the primary terminals are short-circuited, i.e. $V_1 = 0$ and vice-versa.

If in a secondary terminal inductor L is connected, then between the primary terminals we found the capacitance $C = L/s^2$, and if we connect capacitance 'C' in secondary terminals, then between the primary terminals we found inductance $L = s^2 C$. If we connect an impedance Z in secondary terminals, then we find an impedance s^2/Z in primary terminals.

When impedance Z is connected in series, secondary terminals are equivalent to an impedance s^2/Z in parallel to the primary and vice-versa as in Fig. 8.3.

It is known that every inductor has its own resistance and capacitance as well as inductance. If we take the iron-core inductor as in Fig. 8.4a, it can be equivalently modelled as in Fig. 8.4b in which R_1 is the resistance of the inductor wire and C is

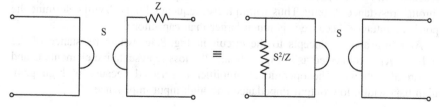

Fig. 8.3 Impedance in series in one pair is equivalent to parallel in another pair.

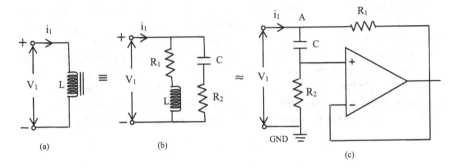

Fig. 8.4 a Simple iron-core inductor. **b** Equivalent model of iron-core inductor. **c** Simulated circuit using the gyrator $L = R_1 R_2 C$. Figure redrawn from popular electronics, July 1977 by B. T. Morrison

the capacitance of the inductor. Resistance R_2 and capacitance C represents losses in iron-core inductor. When DC voltage is applied, there are no core losses, but for higher frequencies core losses are more and increases with increasing frequency; this allows the current flow through resistance R_2 as the capacitor reactance decreases.

By using the passive elements capacitors and resistors with amplifier circuit can synthesize an inductor as in Fig. 8.4c.

A simple *"port admittance"* model would analyse the inductor. Figure 8.4b is the port of inductor where current i_1 flows when input voltage V_1 is applied. When admittance of the port is zero, then current will not flow through it. This means that element is open or perfect insulator, but when element has infinite admittance then infinite current will flow. This means it is short circuit or perfect conductor. So port admittance is the ratio of current flowing into the port to the voltage across the port.

It is known that the *admittance* of inductor L is infinite i.e. short circuited when DC signal is applied. No current will flow through capacitor 'C', it will be as good as open circuited (Fig. 8.4b), so capacitance behaves as open circuit, and no current will flow through R_2.

When infinite frequency signal is applied, L behaves as open circuit. (Inductive reactance $X_L = 2\pi fL$, where f is the signal frequency and L is value of inductor) so as if R_1 is not in the circuit. However, capacitor C is short circuit (Capacitive reactance $X_C = 1/2\pi fC$, where C is the value of capacitor), and entire current flows through resistance R_2 only. Thus within these frequency limits, L will determine the port's admittance, because it is much larger than capacitor C.

Applying above concepts to the circuit in Fig. 8.4c, where resistance of the inductor R_1 is connected as in Fig. 8.4c and the loss representative elements C and R_2 are also shown. The operational amplifier is selected because of high gain, high bandwidth, low output impedance and high input impedance.

8.3 Working of Gyrator

When DC voltage is applied to the input, capacitor C blocks DC, and thus, voltage at non-inverting terminal of Op-Amp is zero. The output of the Op-Amp is at ground potential. So R_1 is virtually connected across the port, i.e. point A and ground. Thus the input current i_1 will flow through R_1 only.

When infinite frequency signal is applied, capacitor C is short-circuited, and thus, the input voltage V_1 reached to non-inverting terminal of Op-Amp. Op-Amp has unity gain, so there will be no voltage drop across R_1 (as if R_1 is removed from the circuit) and the admittance path is through resistance R_2 to ground.

Hence this satisfies the conditions noted in equivalent port model of inductor for low and very high frequencies.

Hence we can say that when frequency increases from lower limit to the higher, the lesser and lesser voltage drop will appear across resistance R_1 causes less port admittance till resistance R_2's effect comes into play. In this way the reactive characteristics of capacitor have been inverted or gyrated, so that port behaves as inductor. The equivalent port inductance is expressed in terms of R_1, R_2 and C as

$$L = R_1 R_2 C$$

For the signal of frequencies between DC to very high, the R_2 and C act as high pass filter because while increasing the frequency the gain of the circuit increases gradually by reducing the voltage drop across R_1, thus port admittance is less.

Practically while selecting the components R_1 should be kept as small as possible and R_2 should be large as possible. Actually R_2 should be at least 100 times greater than R_1 but not so large. For practical purpose R_1 should be around 1 KΩ and R_2 should have the range between 10 K and 1 MΩ. After selecting the values of R_1 and R_2 find the value of capacitor C from the formula

$$C = \frac{L}{R_1 R_2} \tag{8.4}$$

8.4 Advantages of Gyrator

1. No magnetic shielding is necessary due to inductor less circuit.
2. As compared to large value of inductor very small size is required to install. This could increase the compactness of the circuit.
3. Only passive components resistors and capacitor are required.
4. Because of passive components simulated inductor can be fabricated in integrated circuit.
5. Mostly the gyrators are useful for filter design without inductors.

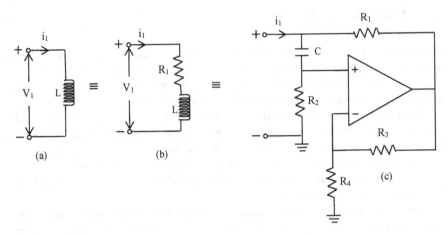

Fig. 8.5 a Air-core inductor. **b** Two-port equivalent model of inductor. **c** Simulated circuit using gyrator. $L = R_1 R_2 C$ ∴ $R_1 = R_3$ and $R_2 = R_4$

8.5 Disadvantages of Gyrator

1. For floating inductors simulation circuit design becomes complex and intricate.
2. Use of active devices generates noise, but this can be reduced by additional circuitry.
3. Low series resistance and high-current inductors are difficult to simulate.
4. Because of active devices frequency limitation is there over simulated inductors.

8.6 Simulation of Air-Core Inductors

Likewise *iron-core or ferrite-core inductors* simulation; air-core inductors can also be simulated. Figure 8.5 shows gyrator circuit using Op-Amp for air-core inductor.

The air-core inductors have no core like iron or ferrite, so core loss is not there. Therefore there is no parallel resistance in the equivalent two port network but for obtaining the proper gain of the Op-Amp two additional R_3 and R_4 resistors are connected. When $R_3 = R_1$ and $R_4 = R_2$ are selected, the circuit will work properly without oscillations otherwise circuit will be destabilized and oscillations will be generated. In this circuit $L = R_1 R_2 C$ with $R_3 = R_1$ and $R_4 = R_2$.

In *gyrator* and *frequency-dependent negative resistance* (FDNR) all the values of elements or passive components are very sensitive; therefore, the accuracy in value has prime importance. Change in the third decimal value or tolerance will change the Q value of the circuit and hence the value of inductor or FDNR will change. Ultimately this may change the cut-off frequency of the filter.

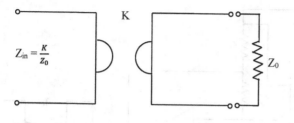

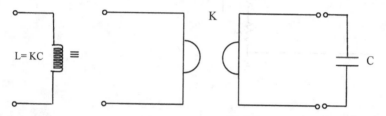

Fig. 8.6 a Lossless two-port network inversion of load impedance. **b** Lossless two-port network simulation of inductor

The gyrator is a lossless two-port circuit which invests load impedance as in Fig. 8.6a, b. When this circuit used with high value Q capacitor, then it simulates the characteristics of an inductor high Q value.

Where $K = \frac{R_1 R_3 R_5}{R_2}$ is a constant.

In any case gyrator requires two Op-Amps and five impedances as in Fig. 8.7. There are two types of gyrators

1. Grounded
2. Floating.

In case of grounded gyrator one end of the element is grounded, and in floating gyrator both the ends of elements are floating and can be used as floating inductor.

In passive LC filter floating gyrators are used. The resistor R_5 is common for opposite ends which simulates floating inductor (Fig. 8.8).

Another way of *simulating floating inductor* is to use the frequency-dependent negative resistor (FDNR) means the value of resistance decreases with increasing frequency of course apparent value.

The passive low pass LC filter is shown in Fig. 8.11. In Fig. 8.7b, when Z_1 and Z_3 replaced by a capacitor C_1 and C_2, then gyrator becomes frequency-dependent negative resistance (FDNR) circuit. We know that the transfer function for basic impedance converter for Fig. 8.7 is

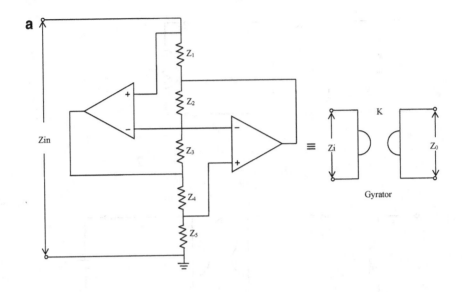

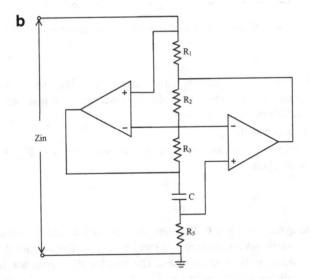

Fig. 8.7 **a** Gyrator circuit inversion of load impedance. **b** Gyrator circuit—RC implementation

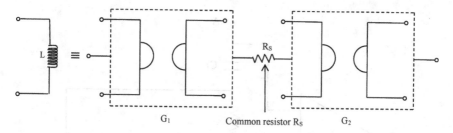

Fig. 8.8 Floating inductor $L = \frac{R_1 R_3 C}{R_2} R_5$

$$Z_{in} = \frac{Z_1 Z_3 Z_5}{Z_2 Z_4} \tag{8.5}$$

If Z_1 and Z_3 are replaced by capacitors and Z_2, Z_4 and Z_5 as resistors, then

$$Z_1 = Z_3 = \frac{1}{j\omega C} = \frac{1}{SC} \quad \because S = j\omega \tag{8.6}$$

Substituting value of Z_1 and Z_3 in Eq. (8.5), we get

$$
\begin{aligned}
Z_{in} &= \frac{1}{S^2 C^2} \frac{R_5}{R_2 R_4} \\
&= \frac{1}{j^2 \omega^2} \frac{R_5}{R_2 R_4 C^2} \\
&= -\frac{1}{\omega^2} \frac{R_5}{C^2 R_2 R_4} \quad \because j = \sqrt{-1} \text{ or FDNR} = Z_{in} = -\frac{1}{\omega^2} \frac{R_5}{C^2 R_2 R_4}
\end{aligned}
\tag{8.7}
$$

8.7 Frequency-Dependent Negative Resistance (FDNR) or D-element

The *frequency-dependent negative resistance* circuit (FDNR) is an active element exhibits real negative resistance that behaves like unusual capacitor. The symbol for FDNR is shown in Fig. 8.9. The FDNR is also called as *D-element*.

Fig. 8.9 Symbol for FDNR

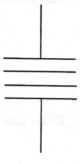

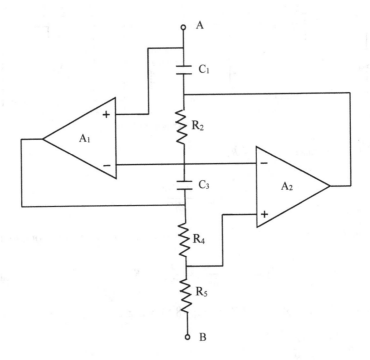

Fig. 8.10 FDNR circuit

The FDNR circuit is shown in Fig. 8.10.

For example if $R_2 = R_4 = R_5 = 1\ \Omega$ and capacitors $C_1 = C_3 = 1\ F$, then the circuit FDNR behaves like a negative resistance of $-1\ \Omega$.

Low pass LC filter using gyrator

In *low pass LC filter*, this FDNR, i.e. impedance Z_{in} of two-port network, is useful to simulate the floating inductor.

The inductive reactance for inductor L_1 and L_2 is

$$X_{L_1} = j\omega L_1 = SL_1 \text{ and} \tag{8.8}$$

$$X_{L_2} = j\omega L_2 = SL_2 \tag{8.9}$$

The capacitive reactance for capacitors C_1 and C_2 is

$$X_{C_1} = \frac{1}{j\omega C_1} = \frac{1}{SC_1} \text{ and} \tag{8.10}$$

$$X_{C_2} = \frac{1}{j\omega C_2} = \frac{1}{SC_2} (\because S = j\omega) \tag{8.11}$$

Divide each element by S, we get

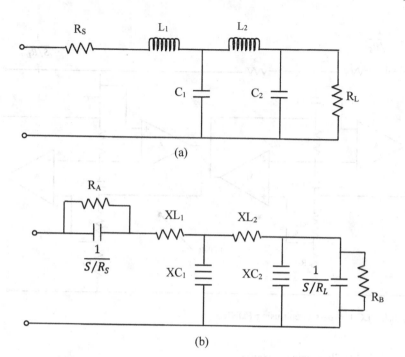

Fig. 8.11 Simulated LC filter using FDNR

$$X_{L_1} = L_1 \text{ and } X_{L_2} = L_2; \frac{R_S}{S} = \frac{1}{S/R_S}$$

$$X_{C_1} = \frac{1}{S^2 C_1} \text{ and } X_{C_2} = \frac{1}{S^2 C_2}; \frac{R_L}{S} = \frac{1}{S/R_L}$$

X_{L_1} and X_{L_2} are impedances replaced by resistors and X_{C_1} and X_{C_2} by FDNR (Fig. 8.11). The R_S and R_L are replaced by capacitors, which are providing bias. Figure 8.11b will be redrawn as Fig. 8.12 with FDNR. R_S is the source resistance and R_L is the load resistance. In LC circuits, while using FDNR transformation capacitors are replaced by FDNR, inductance L by resistor R and resistance R by capacitor C (Fig. 8.12).

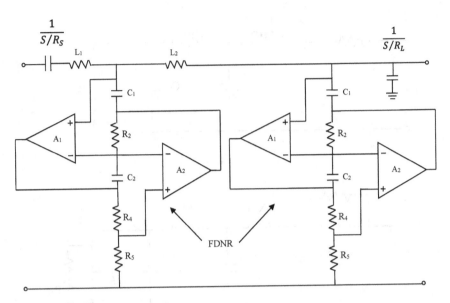

Fig. 8.12 LC low pass circuit using FDNR

High pass LC filter using gyrator

Similarly the high pass filters can be designed using FDNR. But in such case complexity in the design increases, whereas instead of FDNR if gyrator is used the design become simple. So in gyrator circuit gyrator itself is used to replace the inductor. Consider the following high pass filter where capacitors and inductors are used as passive elements (Fig. 8.13).

The circuit of *LC high pass filter* in Fig. 8.13a can be replaced by gyrator without inductor. The value of inductor L would be

$$L = \frac{R_1 R_3 R_5}{R_2} C \qquad (8.12)$$

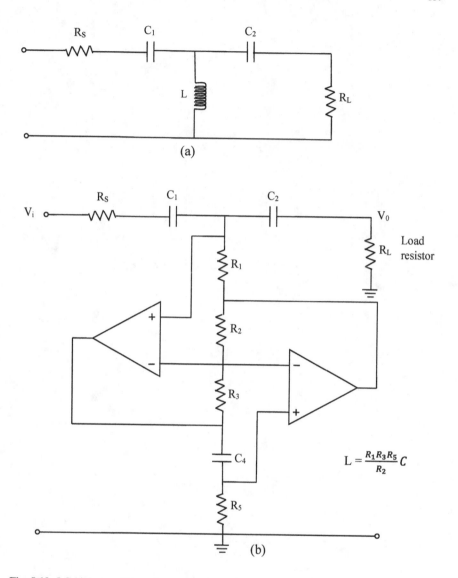

Fig. 8.13 LC high pass filter using gyrator

Very simple concept of ideal gyrator is as follows

1. When load resistance R_L is connected at the output of the gyrator, then its input port behaves as linear resistor with resistance $\frac{1}{K^2 R_L}$ where K is called *gyrator conductance*.

2. When output of a gyrator have capacitor C as a load, i.e. output port is terminated by capacitor C, then its input port behaves as an inductor L, where $L = \frac{C}{K^2}$.

Chapter 9
Noise in Operational Amplifier

This chapter describes noise in operational amplifier. Physically it is the differential amplifier therefore the noise associated with differential amplifier is the noise generated in operational amplifier. While designing the circuit of any amplifier the noise generated due to various components in amplifier circuit cannot be neglected especially when weak signal amplification is required. The noise in input signal is not the only source of noise but fluctuations in a supply voltage or current, loose connection, Brownian motion of electrons in conductor and the transportation of the charges near the junction of semiconductor etc., are the factors responsible for generation of electric noise. Thus while designing the amplifier the electronics engineer or designer has to take into account the various noise sources and how to suppress them. There are various types of noise, viz. Johnson, Schottky, low frequency, popcorn, thermal, shot, etc. In this chapter the classification of noise and their sources are reported. Generally the classification is made as interference and inherent noises. Theory of various types of noise and the noise generated in resistors and capacitors are discussed. Similarly operational amplifier noise model and its analysis in inverting and non-inverting mode have been reported. The equivalent noise models and calculation of total noise from them along with noise figure are given in this chapter.

9.1 Introduction

Basically operational amplifier is a differential amplifier. So the noises associated with differential amplifier are noise generated in operational amplifier. In amplifiers noise has prime importance; it cannot be neglected because the low-intensity input signals having small magnitude of noise will get amplified along with the signal.

© The Author(s), under exclusive license to Springer Nature Singapore Pte Ltd. 2022
S. Yawale and S. Yawale, *Operational Amplifier*,
https://doi.org/10.1007/978-981-16-4185-5_9

So to suppress the noise special modification in the amplifier circuit design is necessary. Not only the noise in the input signal disturbs the working of the amplifier but many more sources of noise such as fluctuations in supply voltage or current, loose connection, Brownian motion of electrons in conductors, etc., are responsible factors to generate electrical noise. The amplifiers are very much sensitive to the electrical noise. This is a significant problem while designing the amplifier circuit.

What is Noise?

The unwanted additional spurious component in the signal is called noise. It is a collection of all unwanted signals coming from external sources as well as the unwanted signal generated internally from components and devices.

9.2 Types of Noise

There are various types of *noise* which are categorized from their sources.

1. Johnson noise
2. Schottky noise
3. Low-frequency (1/f) noise
4. Burst or popcorn noise
5. Thermal noise
6. Shot noise
7. Flicker noise
8. Avalanche noise
9. Spot noise
10. Pink noise

9.3 Classification of Noise

Basically the noise is classified as *interference* and *inherent* noise. These two classified noises are mostly associated with any electronics circuits, to know the sources of such noises and the treatment for suppression has to be considered. Hence the classification of noise which includes various types of noises is given below.

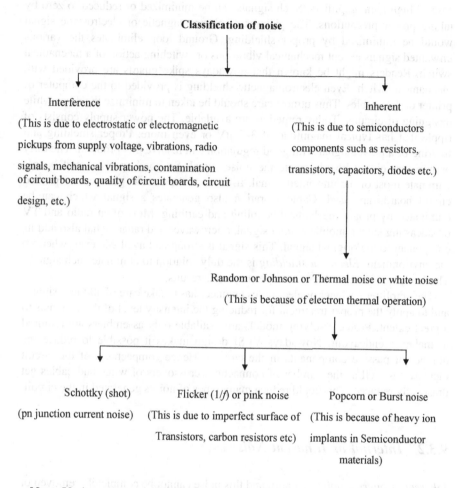

Note: Various sources of the noise created different types of noise. The *internal noise* and *external noise* sources are different.

9.3.1 External or Interference Noise

The external noise sources such as electromagnetic signal or electrostatic pickups vibration, loose connection, sparking, voltage spikes, action of mechanical switch, switching of reactive circuits, vibration of circuit components, contamination of circuit boards, circuit design, crossovers of wires and buses, etc. are the sources of interference noise. This would cause the generation of noise signal which ultimately added to the main signal component and disrupt the working of the amplifier. These unwanted signals get added and amplified to become it appreciable at the output in case of high gain amplifiers. Such signals can be minimized or reduced to zero by taking proper precautions. The pickups of electromagnetic or electrostatic signal would be minimized by proper shielding. Ground loop eliminates the various unwanted signals except mechanical vibrations or switching action of a mechanical switch. Readers might be known that nowadays spike guards are provided with mechanical switch. Even electromagnetic shielding is provided to the computer or printer or data cables. Thus utmost care should be taken to minimize the noise while designing of circuits. Today remedies are available. The power supply consists of ripple that has typical magnitude of 3–5 µV or even more. Proper shielding and filtering of a power signal with good regulation can resolve this problem. The dirty or moisture circuit boards create the noise signal. So good epoxy circuit boards eliminate most of the unwanted signal. In most of the circuits glass epoxy printed circuit boards are used. Cable vibration also generates a signal which can be eliminated by proper mechanical coupling and earthing. Most often radio and TV broadcasting signal, mobile 4G/5G signal, microwaves and radar signal also add its component to the original signal. This signal is strong and available everywhere in the environment. *Electrical shielding* is the only solution to eliminate such signals. The low-capacitance cables can give the better results.

The amplifier designer or electronics engineer has to take care of all these signals and to apply the proper treatment for reducing the intensity level of these signals to a great extent. Problem solving models are available only assemblers are required to make an endeavour. Nowadays VLSI design makes it possible to reduce the number of passive components in the circuits. Hence compactness of the circuit increases as well as the number of connection, crossovers of wires and cables get drastically reduced. This could reduce many types of noises generated in the circuit.

9.3.2 Internal or Inherent Noise

Inherent or internal noise is inborn and this noise cannot be completely removed or eliminated. The interference noise can be eliminated by making various provisions in the amplifier circuit while designing. This noise arises in semiconductor components such as resistors, diodes, transistors and FETs. This noise is called random noise also. The electron movement or transportation from one region to other under constant potential difference may get disturbed due to an increase in the temperature that generates the noise called *thermal noise* or *Johnson noise*. This random or Johnson noise is categorized into three as

1. Schottky or shot noise
2. Popcorn noise
3. Flicker noise or 1/*f* noise

It is known that the theory of noise is well described by Johnson in 1928. Interested readers can read the paper of Johnson for detail theory of noise. The thermal noise is proportional to temperature. In a thermal noise, the power spectral density, $p(f)$ is given by

$$p(f) = KT \qquad (9.1)$$

where T is the temperature of the conductor in Kelvins and K is the Boltzmann constant in J/K. The *power spectral density* $p(f)$ is frequency- and temperature-dependent; hence it is expressed in f/Hz. For a particular bandwidth, Δf (frequencies $f_{max} - f_{min}$) the noise power is expressed as

$$N = p(f)\Delta f$$
$$= KT\Delta f \text{(watts)} \qquad (9.2)$$

Generally the thermal noise generated in a resistor R has the equivalent circuit as in Fig. 9.1.

As per the maximum power transfer theorem when source impedance is matched with the load resistance, maximum power will be delivered to the load. Thus power $p(f)$ is

$$p(f) = \frac{e_{nr}^2}{4R} \qquad (9.3)$$

Equating Eqs. (9.2) and (9.3), we get

$$\frac{e_{nr}^2}{4R} = KT\Delta f$$

Or

$$e_{nr}^2 = 4KTR\Delta f$$

Fig. 9.1 a Noiseless resistor b equivalent circuit of general purpose resistor

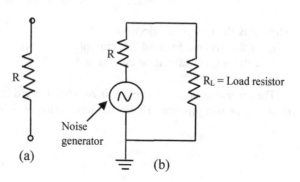

(a)

R Noise generator R_L = Load resistor

(b)

$$e_{nr} = \sqrt{4KTR\Delta f} \qquad (9.4)$$

where e_{nr} is the thermal noise in (*RMS*) volts. The thermal noise current is given by

$$I_n = \sqrt{\frac{4kT\Delta f}{R}} \text{ (rms)} \qquad (9.5)$$

Equations (9.4) and (9.5) indicate that for particular noise bandwidth, the thermal noise voltage e_{nr} is proportional to its resistance and temperature and current is inversely proportional to resistor R. The Johnson noise could be reduced to minimum by selecting small value of resistor at the input of high gain amplifier. Normally this semiconductor noise is very small hence it has less importance. For a resistance of mega ohms at room temperature, this noise is of the order of μV, obviously the noise current is in 1 Pico ampere range. This noise is also called *white noise*. If resistances are in series noise voltage is added. Actually noise current is more important in parallel circuits.

9.3.2.1 Schottky Noise

Schottky noise is generated by the random fluctuations of a motion of electrons. Generally the current flows through a base–emitter junction of transistor constitute a random motion of charges due to variations in the input applied voltage. The energy barrier created due to electrons gets crossed when they acquire sufficient energy and suddenly their potential energy is converted into kinetic energy after crossing the potential barrier. The electrons get shoots across the energy barrier. This constitutes a noise called *shot noise* or *Schottky noise*. In a high gain amplifier or operational amplifier, the bipolar junction transistors at the input stage produce Schottky noise because of input current flowing through base–emitter junction. The shot noise is totally dependent on the flow of charge carriers. If flow of charge carriers continues then shot noise generated, if it stops, the noise stops. This is always associated with current flow hence *RMS* value of Schottky noise current is

$$I_{\text{Schottky}} = \sqrt{(2eI_{dc} + 4eI_0)\Delta f} \qquad (9.6)$$

where e is the charge on electrons,
     I_{dc} is the average forward DC current,
     I_0 is the reverse saturation current and
     Δf is the bandwidth.
     The reverse saturation current I_0 is zero for forward bias base–emitter junction of transistor or p–n junction. Hence the *RMS* value of Schottky noise voltage is

$$V_{schottky} = KT\sqrt{\frac{2\Delta f}{eI_{dc}}} \tag{9.7}$$

At room temperature, if $I_{dc} = 1$ mA, the Schottky noise voltage will be of nV over 20 kHz bandwidth.

9.3.2.2 Flicker Noise or (1/f) Noise

This type of noise is generated at low frequencies below 100 Hz. Generally this noise carries with DC component. It is present in both passive and active components. It is also known as *pink noise*. The imperfect crystalline structure in semiconductors is the main cause of creation of flicker noise. This noise increases with reduction in frequency; therefore it is called *1/f noise* also. For different frequencies this noise contribution is different. Mostly carbon composition resistors contribute this type of noise dominating thermal and shot noise.

The *RMS* flicker noise is given by

$$e_{flicker} = A\sqrt{\log\frac{f_{max}}{f_{min}}} \tag{9.8}$$

where A is the proportionality constant at 1 Hz and f_{max} and f_{min} are the maximum and minimum frequencies in Hz, respectively.

In high gain amplifiers such as operational amplifiers 1/f noise can be reduced by reducing the power consumption.

9.3.2.3 Popcorn or Burst Noise

Popcorn noise consists of sudden step like transitions or pulses of random duration between two or more levels. The intensity of this noise in amplifiers is 100 μV maximum. The main cause of this type of noise generation is crystal imperfection, random trapping and release of charge carriers at the surface of thin films, heavy ion implantation, defect sites in bulk semiconductor, surface contamination, etc. This noise makes a popping sound like popcorn noise. In BJTs and MOS devices or ICs such noise is detected. Good quality and perfect manufacturing process of BJTs or ICs can reduce this noise.

This noise has many name such as *Random telegraph (RTN), bistable, impulse* or *random telegraph signal noise (RTS)*.

This mostly occurs in semiconductors having ultrathin gate oxide films.

9.3.2.4   Avalanche Noise

As its name suggests, *avalanche noise* is detected in reverse bias *p–n* junctions, e.g. Zener diodes. In Zener diode reverse bias electric field creates thick depletion region due to accumulation of opposite charges. These electrons have enough kinetic energy to collide with the atoms of crystal lattice generates election hole pair. These collisions produce random current pulses or spikes of intense magnitudes generate avalanche noise. Avoidance of Zener diodes while designing the circuit can eliminate this noise.

9.4   Noise Contribution in Operational Amplifier

Before discussing the noise contribution in operational amplifier it is necessary to discuss the noise generated due to passive elements such as resistors and capacitors because these elements are often used in any electronic circuitry rather it would not be possible to design any electronics circuit without these external components.

9.4.1   Noise Model of Resistor

When current passes through resistor an electronic noise is generated by the thermal agitation of electrons. Thermal noise is generated in resistor without application of applied voltage. This noise signal is weak but becomes sensitive in many instruments. For example in high gain amplifier already the input signal is weak and if such noise signal is added in that signal, it will get amplified along with the main signal. Mostly the thermal noise generated in a resistor cannot be neglected. The resistor noise model is shown in Fig. 9.2 where noiseless resistor is connected in series for voltage source and in parallel for current source.

If multiple noise sources are present in the circuit then the signal should be combined properly to obtain overall noise signal. The thermal or the Johnson noise voltage is given by

$$e_{nr}^2 = 4KTR\Delta f \tag{9.9}$$

and noise current is given by

$$i_{rms}^2 = \frac{4KT}{R}\Delta f \tag{9.10}$$

When two resistors are connected in series the sum of its noise voltages are independent (Fig. 9.3).

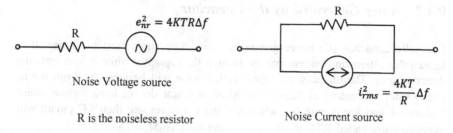

Fig. 9.2 Resistor noise models

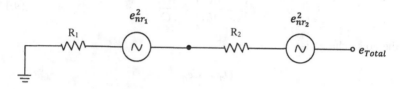

Fig. 9.3 Combination noise model for R_1 and R_2

Thus

The average RMS voltage across the two resistors will be[1]

$$e_{Total}^2 = (e_{nr_1} + e_{nr_2})^2 \tag{9.11}$$

$$= e_{nr_1}^2 + 2e_{nr_1}e_{nr_2} + e_{nr_2}^2$$

$$\approx e_{nr_1}^2 + e_{nr_2}^2 \quad (\because \ 2e_{nr_1}e_{nr_2} \approx 0)$$

Or

$$\approx 4KTR_1\Delta f + 4KTR_2\Delta f \quad (\because e_{nr_1}^2 = 4KTR_1\Delta f; e_{nr_2}^2 = 4KTR_2\Delta f)$$

$$e_{Total}^2 = 4KT(R_1 + R_2)\Delta f \tag{9.12}$$

This equation is true for current sources also as in voltage sources.
In general for more number of series resistors, we can write

$$e_{Total}^2 = 4KT(R_1 + R_2 + R_3 + \cdots)\Delta f \tag{9.13}$$

---

[1]Noise analysis in Operational amplifier Circuits, Application Report, (2007), SLVA043B Texas Instruments.

9.4.2 Noise Generated by the Capacitor

Generally capacitor (C) never generates noise, they suppressed the noise. It is known that alternating current passes through the capacitor while it interrupts the direct current. The unnecessary signal called noise will be removed from the ac mains using capacitors as filters. The ideal capacitor do not have thermal noise because of loss less nature but when it is used with resistor then RC circuit will generate noise called KTC noise whose bandwidth is $\Delta f = \frac{1}{4RC}$.

Thus *RMS* voltage generated by the RC filter is

$$e_m^2 = \frac{4KRT}{4RC} = \frac{KT}{C} \tag{9.14}$$

and the noise charge is

$$Q_n = Ce_m$$

$$= C\sqrt{\frac{KT}{C}} \quad \left(\because \frac{Q}{C} = V\right)$$

Or

$$Q_n^2 = KTC \tag{9.15}$$

Hence capacitor noise is called KTC noise. This noise is generated entirely due to resistor thus the temperature of the resistor itself is considered.

9.4.3 Operational Amplifier Noise Model

Generally operational amplifier is used in inverting, non-inverting and differential mode configuration. It is necessary that the contribution of each component should be taken into consideration for calculation of total noise. So series resistor, feedback resistor, current flowing through inverting and non-inverting terminals and Op-Amp noise contribution are evaluated and then sum of all these gives total noise.

Nice explanation and evaluation of various noise parameters have been discussed by Tim J. Sobering[2] in his technical note. Also Levis Smith and D. H. Sheingold[3] have explained the noise in operational amplifier circuits.

---

[2]Tim J. Sobering, Op Amp Noise Analysis, Technote 5, May 1999 revised 11/19/02 by Tim J. Sobering, SDE Consulting, © 1999 Tim J. Sobering.
[3]Levis Smith and D. H. Sheingold, Noise and Operational amplifier circuits, Analog Dialogue 3(1) March 1969.

9.4.3.1 Noise Analysis in Inverting Mode

The fundamental circuit of inverting mode operational amplifier is described in Chap. 2. The well-known circuit of inverting mode Op-Amp is given in Fig. 9.4.

The equivalent noise model for above circuit configuration is given in Fig. 9.5, where noiseless (ideal Op-Amp) operational amplifier is used along with noise sources. Where i_+ and i_- are the input current generated due to noise in non-inverting and inverting inputs of Op-Amp, respectively. e_n is the amplifier noise voltage when source resistance is zero. i_+ and i_- noises contribute when source resistance is nonzero. All these sources have thermal or Johnson, $1/f$ shot and

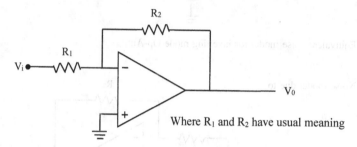

Where R_1 and R_2 have usual meaning

Fig. 9.4 Inverting Op-Amp

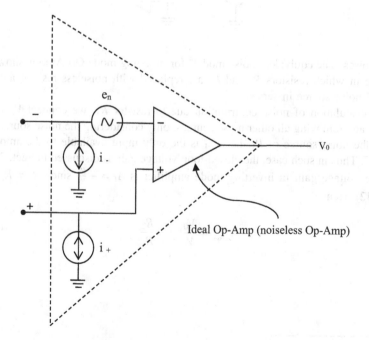

Ideal Op-Amp (noiseless Op-Amp)

Fig. 9.5 Op-Amp noise model

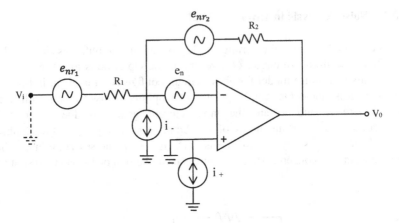

Fig. 9.6 Equivalent noise model for inverting mode Op-Amp

Fig. 9.7 Noise model due to R_1

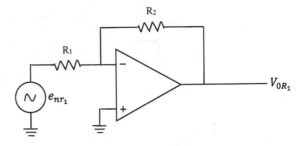

other noises. The equivalent noise model[4] for inverting mode Op-Amp is shown in Fig. 9.6 in which resistors R_1 and R_2 are replaced with noiseless resistor and the thermal noise source in series.

For calculation of noise contribution due to resistor R_1, we connect V_i to the ground and removing all other noise sources, only considering the noise source due to R_1, the noise source of R_1, i.e. e_{nr1} is the only input available to the amplifier Fig. 9.7. Thus in such case, the closed-loop voltage gain is discussed in Sect. 2.5.1, and the voltage gain in inverting mode amplifier is $A_f = -\frac{R_2}{R_1}$ since $R_2 = R_f$ (see Eq. 2.42). Hence

$$A_{f_{noise}} = \frac{V_0}{V_i} = -\frac{R_2}{R_1} \tag{9.16}$$

Or

---

[4]Bruce Carter, Op Amp for Everyone, Elsevier Pub. Chap. 10 (Texas Instruments)

$$V_0 = -V_i \frac{R_2}{R_1}$$

Since the inherent input voltage source due to resistor R_1 (noise voltage) is e_{nr1}, hence $V_i = e_{nr1}$. So the output voltage represented by V_{oR1} is given by

$$V_{OR_1} = -e_{nr_1} \frac{R_2}{R_1} \quad \left(\because \ V_i = e_{nr_1} = \sqrt{4KTR_1\Delta f} \right)$$

$$V_{OR_1} = -\sqrt{4KTR_1\Delta f} \frac{R_2}{R_1} \tag{9.17}$$

Equation (9.17) suggests the output noise voltage due to resistor R_1 in inverting mode Op-Amp.

Similarly noise contribution due to feedback resistor R_2 alone can be calculated using following Fig. 9.8, where all other noise sources are removed and only noise source due to R_2 (i.e. e_{nr_2}) is considered.

Applying Kirchhoff's current law (KCL) at the node A, we get

$$i_1 + i_2 = 0$$

$$\frac{V_a}{R_1} + \frac{V_{0R_2} - e_{nr_2} + V_a}{R_2} = 0 \tag{9.18}$$

On solving this Eq. (9.18), we get

$$V_{0R_2} = e_{nr_2} \frac{\beta A}{1 + \beta A} \quad \left(\because \ \beta = \frac{R_1}{R_1 + R_2} \text{ and } A = \frac{V_0}{V_a} \right) \tag{9.19}$$

The open-loop gain A of an ideal Op-Amp is infinite ($A \rightarrow \infty$) and $\frac{\beta A}{1+\beta A} = 1$. Thus

$$V_{0R_2} = e_{nr_2}$$

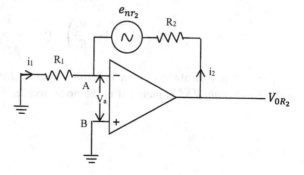

Fig. 9.8 Noise model due to R_2

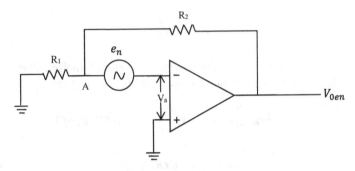

Fig. 9.9 Noise mode due to contribution of e_n in case of inverting mode Op-Amp

$$V_{0_{R_2}} = \sqrt{4KTR_2\Delta f} \quad \left(\because\ e_{nr_2} = \sqrt{4KTR_2\Delta f} \right) \tag{9.20}$$

This Eq. (9.20) shows that the output V_0 is the noise voltage generated due to R_2 only.

Similarly the contribution due to source e_n can be calculated by keeping only source e_n and removing all other noise sources. Figure 9.9 shows the noise model for e_n only.

Applying KCL at the node A, we get

$$\frac{-e_n + V_a}{R_1} + \frac{V_0 - e_n + V_a}{R_2} = 0 \tag{9.21}$$

$$- e_n\left(\frac{1}{R_1} + \frac{1}{R_2}\right) + V_0\left(\frac{1}{R_2} + \frac{1}{AR_1} + \frac{1}{AR_2}\right)$$

$$= 0 \quad \left(\because\ A = \frac{V_0}{V_a}\ \text{and}\ Vo = V_{0_{en}} \right)$$

Or

$$V_{0_{en}} = e_n \frac{A(R_1 + R_2)}{AR_1 + (R_1 + R_2)} \tag{9.22}$$

$$V_{0_{en}} = e_n\left(1 + \frac{R_2}{R_1}\right) \quad (\because A \to \infty) \tag{9.23}$$

The e_n value is always given in terms of *noise density* (nV/$\sqrt{\text{Hz}}$), therefore *noise bandwidth* $\sqrt{\Delta f}$ is included in the noise source.

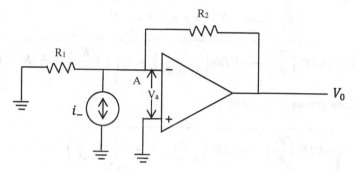

Fig. 9.10 Noise model for i_- contribution

$$\therefore V_{0_{en}} = e_n \sqrt{\Delta f}\left(1 + \frac{R_2}{R_1}\right) \qquad (9.24)$$

This equation represents the output of Op-Amp when noise source e_n is considered.

The noise contribution due to noise current i_- can be calculated by keeping the current source i_- only and removing other noise sources. Figure 9.10 shows the noise model for i_- contribution.

Applying KCL at point A, we get

$$\frac{V_a}{R_1} + \frac{V_0 + V_a}{R_2} - i_- = 0 \qquad (9.25)$$

On solving this Eq. (9.25), we get

$$V_0 = i_-(R_2) \qquad (9.26)$$

i_- is typically mentioned in noise density $(PA/\sqrt{Hz})$, the noise bandwidth $\sqrt{\Delta f}$ is included so

$$V_0 = i_-\sqrt{\Delta f}(R_2) \qquad (9.27)$$

The noise contribution due to i_+ is zero because the non-inverting terminal of Op-Amp is grounded hence it is shorted. Therefore i_+ contribution is zero (refer to Fig. 9.6).

Hence the total noise can be computed by combining the resistor noise and Op-Amp noise. By using Eqs. (9.17, 9.20, 9.24 and 9.27) the thermal noise for inverting mode Op-Amp is

$$V_{0rms} = V_{0R_1} + V_{0R_2} + V_{0en} + V_{0i-} \tag{9.28}$$

$$= \sqrt{\Delta f} \sqrt{4KTR_1 \left(\frac{R_2}{R_1}\right)^2 + 4KTR_2 \left(\frac{R_1}{R_1 + R_2}\right)^2 + e_n^2 \left(1 + \frac{R_2}{R_1}\right)^2 + i_-^2 (R_2^2)} \tag{9.29}$$

The noise density is given by

$$\frac{V_{0rms}}{\sqrt{\Delta f}} = \sqrt{4KTR_1 \left(\frac{R_2}{R_1}\right)^2 + 4KTR_2 \left(\frac{R_1}{R_1 + R_2}\right)^2 + e_n^2 \left(1 + \frac{R_2}{R_1}\right)^2 + i_-^2 (R_2^2)} \tag{9.30}$$

This Eq. (9.30) gives total output noise of the inverting mode Op-Amp.

9.4.3.2   Noise Analysis in Non-Inverting Mode Op-Amp

In non-inverting mode Op-Amp we connect resistors R_1 and R_2 as in the case of inverting mode except applying signal to inverting input, signal is applied through non-inverting input and an additional resistance R_3 is connected in non-inverting input through which input signal is applied. This resistance R_3 is generally parallel combination of R_1 and R_2 for bias current cancellation. Figure 9.11 shows the circuit of Op-Amp, where R_3 is connected in non-inverting input.

The noise sources due to R_2 and R_3 are not considered now hence not shown in Fig. 9.11.

To evaluate the noise output voltage due to contribution of resistor R_1, the circuit shown in Fig. 9.11 is very similar to the circuit shown in Fig. 9.7 and noise model due to R_1 is in inverting mode Op-Amp except R_3. Thus the noise output voltage generated is similar to V_{0R_1} in inverting mode. Hence V_{0R_1} in non-inverting mode is

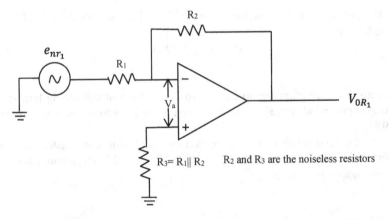

Fig. 9.11 Equivalent noise model when R_3 is connected in non-inverting input of Op-Amp

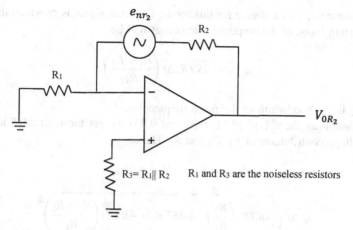

Fig. 9.12 Equivalent noise model due to R_2 in non-inverting mode

$$V_{OR_1} = \sqrt{4KTR_1\Delta f}\left(\frac{R_2}{R_1}\right) \tag{9.31}$$

Similarly the noise voltage generated by feedback resistance R_2 can be evaluated (see Fig. 9.12).

Refer to the Fig. 9.8 in inverting mode Op-Amp, similar noise output voltage will be generated at the output. Hence

$$V_{OR_2} = \sqrt{4KTR_2\Delta f} \tag{9.32}$$

Now to evaluate the noise output voltage due to contribution of resistor R_3. The equivalent noise model for this case is shown in Fig. 9.13.

Fig. 9.13 Equivalent noise model for R_3 in non-inverting mode

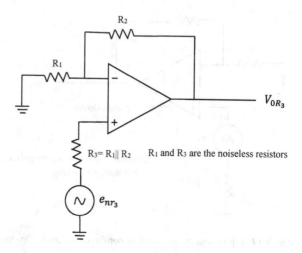

Considering e_{nr_3} as a source for this mode, i.e. input signal is provided through non-inverting input of the amplifier, the output will be

$$V_{OR_3} = \sqrt{4KTR_3\Delta f}\left(\frac{R_1 + R_2}{R_1}\right) \tag{9.33}$$

where Δf is the bandwidth of the noise frequencies.

On combining these Eqs. (9.31, 9.32 and 9.33) we get the total *RMS* thermal noise voltage contribution of R_1, R_2 and R_3. Hence

$$V_{0rms} = V_{OR_1} + V_{OR_2} + V_{OR_3} \tag{9.34}$$

$$= \sqrt{\Delta f}\sqrt{4KTR_1\left(\frac{R_2}{R_1}\right)^2 + 4KTR_2 + 4KTR_3\left(\frac{R_1 + R_2}{R_1}\right)^2} \tag{9.35}$$

On solving this Eq. (9.35), we get

$$V_{0rms} = \sqrt{\Delta f}\sqrt{4KTR_2\left(\frac{R_1 + R_2}{R_1}\right) + 4KTR_3\left(\frac{R_1 + R_2}{R_1}\right)^2} \tag{9.36}$$

Above equation gives the noise output voltage of Op-Amp due to contribution of resistors R_1, R_2 and R_3 only.

Let us calculate the contribution of noise sources associated with Op-Amp itself. Refer to the earlier Fig. 9.9 of Op-Amp noise model, in which Op-Amp (IC) noise we have considered in inverting mode. Here we are using non-inverting mode, hence the Op-Amp noise e_n to be consider again. Figure 9.14 shows the equivalent noise model in such case where noiseless Op-Amp (IC) is considered.

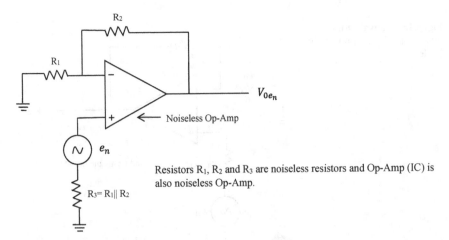

Resistors R_1, R_2 and R_3 are noiseless resistors and Op-Amp (IC) is also noiseless Op-Amp.

Fig. 9.14 Equivalent noise model in non-inverting mode Op-Amp

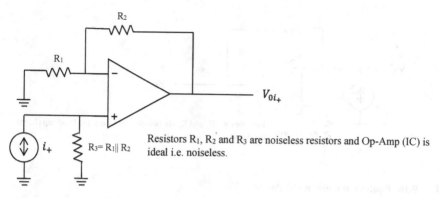

Resistors R_1, R_2 and R_3 are noiseless resistors and Op-Amp (IC) is ideal i.e. noiseless.

Fig. 9.15 Equivalent noise models for i_+

e_n is the noise generated due to Op-Amp itself. In such case the noise output voltage will be

$$V_{0e_n} = \sqrt{e_n^2 \Delta f \left(\frac{R_1 + R_2}{R_1}\right)^2} \tag{9.37}$$

Besides this voltage source e_n, the i_+ and i_- input current noises in non-inverting and inverting inputs of Op-Amp should be considered. In this case the equivalent noise mode is shown in Fig. 9.15.

In this case the noise output voltage due to i_+ current source is given by

$$V_{0i_+} = \sqrt{i_+^2 R_3^2 \Delta f \left(\frac{R_1 + R_2}{R_1}\right)^2} \tag{9.38}$$

Similarly the contribution occurs due to i_- current source can be calculated. Figure 9.16 shows the equivalent noise model of Op-Amp in non-inverting mode contribution due to i_- current source.

The noise output voltage contributed because of current source i_- will be

$$V_{0i_-} = \sqrt{i_-^2 R_2^2 \Delta f} \tag{9.39}$$

On combining the contributions of e_n, i_+ and i_-, we get noise output voltage of Op-Amp itself

$$V_{0rms} = \sqrt{e_n^2 \left(\frac{R_1 + R_2}{R_1}\right)^2 + i_+^2 (R_3^2) \Delta f \left(\frac{R_1 + R_2}{R_1}\right)^2 + i_-^2 (R_2^2) \Delta f} \tag{9.40}$$

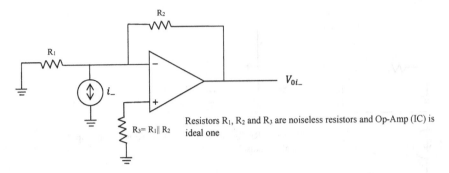

Fig. 9.16 Equivalent noise model due to i_-

The total noise *RMS* voltage generated at the output will be the combination of noise voltages due to resistors R_1, R_2 and R_3 and e_n, i_- and i_+ of Op-Amp itself.

Thus on combining Eqs. (9.36) and (9.40) we get,

V_{0rms} (Total)

$$= \sqrt{\Delta f} \sqrt{4KTR_2 \left(\frac{R_1+R_2}{R_1}\right) + 4KTR_3 \left(\frac{R_1+R_2}{R_1}\right)^2 e_n^2 \left(\frac{R_1+R_2}{R_1}\right)^2 + i_+^2 (R_3^2) \left(\frac{R_1+R_2}{R_1}\right)^2 + i_-^2 (R_2^2)}$$

$$(9.41)$$

Above equation gives the total output noise voltage in non-inverting mode Op-Amp. Equivalent noise bandwidth (ENB) is used to account for extra noise brick-wall frequency limits and determined by the frequency characteristics of the circuit where $(f_H/f_L) = $ ENB, and f_H and f_L are highest and lowest frequencies.

The noise voltage or current can be determined over a frequency band over f_L and f_H. So within these frequency limits of these frequencies white noise and 1/ f noise voltage can be evaluated.

The operational amplifier contains white as well as 1/f noise. Thus in a frequency spectrum where 1/f and white noise are equal that frequency is called noise corner frequency, f_{nc} (Fig. 9.17).

On the similar lines the differential Op-Amp circuit noise can be calculated. All voltages considered or used in equations are root-mean-square (RMS) voltages.

9.5 Noise Figure

Noise figure is a measure of the additional noise contributed by the amplifier over and above that of the source resistance. It can be computed by the following formula.

Fig. 9.17 Input noise versus
frequency plot

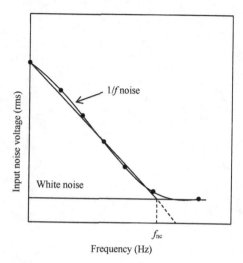

$$\text{Noise figure (NF)} = 10 \log \left[\frac{e_n^2 + i_n^2 R_s^2 + 4KTR_S \Delta f}{4KTR_S \Delta f} \right] \qquad (9.42)$$

where R_S is the optimum source resistance is equal to e_n/i_n, T is the temperature, K is Boltzmann constant and Δf is the bandwidth. The value of R_S may vary with bandwidth. Generally noise figure is expressed in terms of dB.

The noise figure is expressed in terms of actual closed-loop configuration and computed RMS value of noise. Hence

$$\text{N.F.} = 10 \log \frac{\text{Total output noise}}{\text{Source resistor noise}} \qquad (9.43)$$

Or

$$\text{N.F.} = \frac{(\text{Signal to noise ratio at input})}{(\text{Signal to noise ratio at output})}$$

$$= \frac{(S/N)_{\text{input}}}{(S/N)_{\text{output}}} \qquad (9.44)$$

$$\text{N.F.(dB)} = \frac{10 \log (S/N)_{\text{input}}}{10 \log (S/N)_{\text{output}}} \qquad (9.45)$$

The noise figure in Op-Amp depends on input referred voltage, current noise, circuit topology and the value of external components. The negative feedback can improve *S/N* ratio significantly.

Appendix I: Experiments

The students learning using laboratory- based experiments provides new insight into them. The practical skill and the knowledge should go hand in hand. This ultimately gives experience of learning and confidence. The laboratory-based practicals provide some different experience in science students. The learning potential is explored in the students due to laboratory- based experiments. The preparation of experiments, report writing, after performing the laboratory work, conclusions drawn, etc. can enhance the performance of the students. Working with the team and approaching towards the new conclusion gives happiness and more knowledge than teaching to the students. Teaching and learning without practicals or laboratory work in science and engineering and technology is futile. The student should be well acquainted with the laboratory practice. Therefore technology-enhanced learning supports the learning of science. Laboratory learning has a distinctive role in science education. Since long the experimenters are doing science experiments without prior much knowledge of relevant theory. Many times they get success in their experiments. This can only be achieved by learning the laboratory-based work. Without performing the experiment or laboratory work itwould not be possible for the students to draw conclusions. Unless he performs the practical work his thought process is meaningless. Moreover the students get knowledge to handle new and modern equipments. Once he performs the laboratory work it becomes his lifetime attainment, and the practical knowledge he acquired can be imparted to others. In view of this few laboratory exercise/experiments on Op-Amp are described. The relevant theory of the reported experiments is already depicted in respective chapters. Hence students should not find any difficulty while performing the experiments. In all some 17 experiments/laboratory exercise on voltage shunt and series feedback, adder, subtractor, integrator, differentiator, active filters (low, band, high pass and notch), voltage to frequency and frequency to voltage converter, precision reciter, instrumentation amplifier, phase shift, Wien bridge oscillators, astable multivibrator, waveform generator, Schmitt trigger, etc., are given.

The pin configuration of IC 741, 78XX, 79XX is also reported. The operational amplifier requires dual ±15 V regulated power supply. Simple design using one transformer, four diodes, two capacitors and ICs 7815 and 7915 is given in the appendix. Students can easily fabricate this power supply in the small laboratory. This power supply is not costly, and hence students can afford it.

Experiment No. 1

To study the voltage shunt feedback in operational amplifier (Inverting mode Op-Amp).

Aim	To obtain the closed loop voltage gain
Apparatus	Readymade kit of Op-Amp (if available) or breadboard, regulated power supply ±15 V, IC 741, resistors 100 K and 10 K, 10 K potentiometer or preset, connecting wires, hookup wire, oscillator, multimeter, digital voltmeter (DVM).
Formula	closed loop voltage gain

$$A_V = -\frac{R_f}{R_1}$$

where

R_f is the feedback resistor, and

R_1 is the input series resistor

PROCEDURE This experiment can be performed in two ways. (a) Either by changing the values of feedback resistor (R_f) and input resistor (R_1) and keeping input signal voltage fixed or (b) by keeping the R_f and R_1 values fixed and changing amplitude of input signal.

1. In the *first case*—Make the connections as shown in Fig. E.1 by selecting proper values of feedback and input resistors.

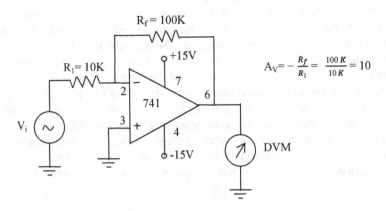

$$A_V = -\frac{R_f}{R_1} = \frac{100\,K}{10\,K} = 10$$

Fig. E.1 Inverting mode Op-Amp

Connect the dual regulated power supply having ± 15 V as shown in Fig. E.1. Connect 10 K potentiometer between pin no. 1 and 5 (of IC 741) with its extreme ends and variable terminal of potentiometer to $-Vcc$. Connect pin no. 2 and 3 (of IC 741) to the ground, and adjust the potentiometer resistance so that the output voltage should be zero. This arrangement is called *offset null*. Keep this position as it is throughout the experiment. Do not change the resistance of the potentiometer. The input signal source V_i may be either DC or AC signal. Apply fixed input signal V_i to the pin no. 2 of IC 741 through resistor R_1. The pin no. 3 should be connected to ground terminal. Preferably make the connections on breadboard so that the circuit can be assembled or de-assembled as per the requirement. Select proper values of feedback resistor such as $R_f = 100$ K (say) and $R_1 = 10$ K (say). Switch on power supply and apply a fixed input signal $V =_i 0.1$ V (say). Note down the output voltage V_0 by DVM in observation Table E.1. For various values of R_f and R_1 calculate theoretical and experimental gain and verify.

Observation Table E.1

Sr. No	Resistors	Input voltage (V)	Output voltage (V)	Voltage gain $A_V = -\frac{V_0}{V_i}$	Theoretical Voltage gain $A_V = -\frac{R_f}{R_1}$
1	$R_f = 100$ K $R_1 = 10$ K				-10
2	$R_f = 470$ K $R_1 = 47$ K				-10
3	$R_f =$ $R_1 =$				
4	$R_f =$ $R_1 =$				

Results and Discussion

Theoretical and experimental gain are verified and found to be identical. Negative output voltage shows the 180° phase shift between input and output voltage.

Second case—Make the connections as shown in Fig. E.1. Connect $R_f = 100$ K and $R_1 = 10$ K. Switch on the power supply. Apply signal of fixed frequency or DC signal to the input terminal, and measure the corresponding output voltage by varying the amplitude of the input signal. Note down the observations in Table E.2. Calculate the voltage gain A_V, and compare the theoretical gain with experimental gain.

Observation Table E.2

$$R_f = 100\,\text{K} \quad R_1 = 10\,\text{K} \quad A_V = -\frac{R_f}{R_1} = \frac{100\,\text{K}}{10\,\text{K}} = 10$$

Sr. No.	Input voltage V_i (V)	Output voltage V_0 (V)	Voltage gain $A_V = -\frac{V_0}{V_i}$
1	0.1		
2	0.2		
3	0.3		
4	0.4		
5	0.5		
6	0.6		
7	0.7		
8	0.8		
9	0.9		

Results and Discussion

Theoretical and experimental gain were compared and were found to be identical. The output voltage is found to be negative which shows that there is 180° phase shift between input and output voltage.

Precautions

1. Dual power supply must be regulated.
2. Apply very small input voltage so that the Op-Amp should not be saturated, because of the restrictions of amplifier that the amplified output should not go beyond supply voltage.
3. Measure all voltages with respect to ground.
4. Use high input impedance voltmeter (either analog or digital) for voltage measurement to avoid the loading.
5. Use proper values of R_f and R_1.
6. For applying the small voltages at the input terminal, a voltage divider network can be used.

Viva Questions

1. What is called differential amplifier?
2. What are the typical characteristics of differential amplifier?
3. Why there is a need of two power supplies in differential amplifier?
4. What do you mean by single-ended and double-ended differential amplifier?
5. Explain working of emitter follower
6. What are the properties of ideal Op-Amp?
7. List few applications of Op-Amp?
8. What do you mean by inverting and non-inverting mode of Op-Amp?
9. What is the gain of operational amplifier in inverting and non-inverting mode?

Experiment No. 2

To study the voltage series feedback in operational amplifier (Non-inverting mode Op-Amp).

Aim   To obtain the closed loop voltage gain

Apparatus   Readymade kit of Op-Amp (if available) or breadboard, regulated power supply ±15V, IC 741, resistors 100 K, 10 K, etc., connecting wires, hookup wires, oscillator and DVM.

Formula   Closed loop voltage gain

$$A_V = 1 + \frac{R_f}{R_1}$$

where

R_f is the feedback resistor, and

R_1 is the input resistor.

Diagram   .

Procedure   The procedure of this experiment is same as that of inverting amplifier experiment no. 1. Instead of applying signal input to pin no. 2 of 741 Op-Amp IC, it is applied to pin no. 3. This experiment can also be performed in two ways as described in inverting amplifier experiment (Fig. E.2).

The observation table is also similar, only theoretical voltage gain $A_V = 1 + \frac{R_f}{R_1}$ should be mentioned, and accordingly the results and discussion should be written.

Fig. E.2 Non-inverting Op-Amp

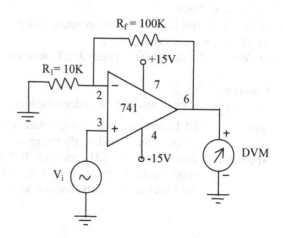

Results and Discussion

Theoretical and experimental gains were compared and were found to be identical. The output voltage is found to be positive which shows that there is no phase shift between input and output voltage. The voltage gain is always greater than one.

Precautions

1. Input impedance of the measuring instrument should be high.
2. Source should have low output impedance.
3. Use 1% tolerance resistors in the circuit.
4. Dual power supply must be regulated.
5. Apply very small input voltages so that the Op-Amp should not be saturated, because of the restriction of the amplifier that the amplified output should not go beyond supply voltage.
6. Measure all voltage with respect to ground.
7. Use high input impedance voltmeter (either analog or digital) for voltage measurement to avoid the loading.
8. Use proper values of R_f and R_1.
9. For applying the small voltages at the input terminal, a voltage divider network can be used.

Viva Questions

1. What is the gain of differential amplifier in non-inverting mode?
2. On which factors the gain of Op-Amp depends?
3. Define input offset voltage and output offset voltage?
4. What is CMRR?
5. Define slew rate?
6. What will be the effect of slew rate on any application?
7. What is PSRR?
8. What do you know about temperature drift?
9. Give few names of Op-Amp IC.
10. What is the difference between JFET input and BJT input stage Op-Amp?

Experiment No. 3

To study operational amplifier as adder/subtractor

Aim   (i) To add the given voltages and verify the result (ii) To subtract the given voltages and verify the result

Apparatus   Breadboard, IC 741, resistors 100 K, 10 K, 1 K (1/4W), power supply ±15 V, rheostats or 1 K (linear) wire-wound potentiometer for variation of inputs, hookup wire, digital multimeter or DVM.

Formula   (i) for adder.

Output voltage $V_0 = -\left(\frac{R_f}{R_a} V_1 + \frac{R_f}{R_b} V_2 + \frac{R_f}{R_c} V_3\right)$.

If $R_a = R_b = R_c = R_f = R$ then

$$V_0 = -(V_1 + V_2 + V_3) \tag{E.1}$$

where V_1, V_2 and V_3 are the input voltages to be added, and V_0 is the output voltage (ii) for subtractor

$$V_0 = -(V_2 - V_1) \tag{E.2}$$

Procedure   (i) Adder or summing and subtractor or difference inverting amplifier - Make the connections as shown in Fig. E.3 on breadboard. Use feedback resistor R_f and input resistors $R_a = R_b = R_c = R_f = R = 10$ K. The input voltages V_1, V_2 and V_3 should be made variable using voltage divider arrangement. Apply separate voltages V_1, V_2 and V_3. Switch on the power supply, apply input voltage to be added/subtracted (as the case may be), and note the corresponding output voltage using DVM. Record the observations in observations Table (E.3), and verify the output (Fig. E.4).

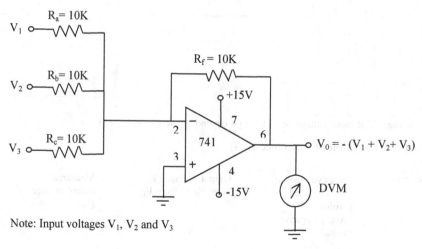

Note: Input voltages V_1, V_2 and V_3

Fig. E.3 a Op-Amp as adder or summing amplifier

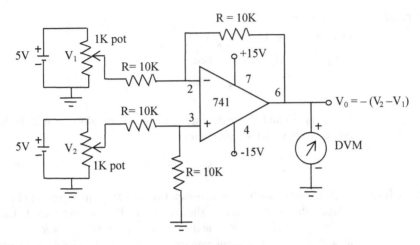

Fig. E.4 b Op-Amp as subtractor or difference amplifier

Observation Table (E.3): (i) for addition

$$R_f = R_a = R_b = R_c = R = 10\,K$$

Sr. No.	Input voltages to be added (volts)			Output voltage (V_0) volts $V_0 = (V_1 + V_2 + V_3)$	Measured output voltage (V_0) V
	V_1	V_2	V_3		
1					
2					
3					
4					
5					
6					

Note The added input voltage should not go beyond the supply voltage (V_{CC})

(ii) For subtraction or difference (E.4)

Sr. No.	Input voltages to be subtracted (volts) Apply negative voltage V_1 (i.e. $-V_1$)		Output voltage (V_0) volts $V_0 = (V_2 - V_1)$	Measured output voltage (V_0) V
	$-V_1$	V_2		
1				
2				
3				

(continued)

(continued)

Sr. No.	Input voltages to be subtracted (volts) Apply negative voltage V_1 (i.e. $-V_1$)		Output voltage (V_0) volts $V_0 = (V_2 - V_1)$	Measured output voltage (V_0) V
	$-V_1$	V_2		
4				
5				
6				

Results and Discussions

In case of adder and subtractor, the measured output voltage is found to be identical as that of output voltage obtained from formula of addition or subtraction. Hence the result is verified.

Precautions

1. Dual power supply must be regulated.
2. Apply very small input voltages so that the Op-Amp should not be saturated, because of the restrictions of amplifier that the amplified output should not go beyond supply voltage.
3. Measure all voltages with respect to ground.
4. Use high input impedance voltmeter (either analog or digital) for voltage measurement to avoid the loading.
5. Use proper values of R_f and R_1.
6. For applying the small voltages at the input terminal, a voltage divider network can be used.
7. Use regulated power supply for input voltages.
8. Input voltage V_1, V_2 and V_3 should have exact values.

Viva Questions

1. Why CMRR should be large in differential amplifier?
2. What do you mean by differential mode gain?
3. Give examples of linear circuits of operational amplifier?
4. Explain adder, summing amplifier and scalar?
5. Define input bias current?

Experiment No. 4

Application of Op-Amp IC 741 as integrator

Aim

    (i)   To study the Op-Amp as integrator
   (ii)   To observe the output waveform when square wave signal is applied at input
 (iii)   Observe the change in output voltage when variable frequency signal is applied at the input.

Apparatus

Breadboard, IC 741, Op-Amp, resistors (1/4 w), capacitor, power supply ± 15V, oscillator (sinusoidal and square waveform generator), hookup wire, *RMS* digital voltmeter, etc.

Formula

$$\text{Output } V_0 = -\int \frac{1}{RC} V_i \mathrm{d}t \qquad (E.3)$$

where V_i is the input signal voltage, R and C are the components connected in circuit as resistor and capacitor, selected as per $T \le RC$ required condition for integration. If V_i is a sinusoidal wave as,

$$V_i = V \sin wt \qquad (E.4)$$

Then $V_o = -\int \frac{1}{RC} V \sin \omega t \mathrm{d}t$

$$= \frac{1}{RC\omega} V \cos \omega t \qquad (E.5)$$

Or

$$V_0 \infty \frac{1}{\omega} \qquad (E.6)$$

Equation (E.6) shows that the amplitude of the output signal is inversely proportional to the frequency of the input signal.

Operating frequency

$$f_a = \frac{1}{2\pi R_f C_f}$$

0 dB Frequency

$$f_b = \frac{1}{2\pi R_1 C_f}$$

Procedure

Make the connections as shown in Fig. (E.5) on bread-board. Select proper values of R_f, R_1 and C_f so that $T \leq R_1C_f$ is satisfied. By choosing proper value of C_f, calculate value of resistor R_1 from $f_b = \frac{1}{2\pi R_1 C_f}$, R_f is the resistor having high value, which provides feedback. f_b is the frequency below which the circuit can work as practical integrator (Fig. E.6). Connect oscillator at the input, and measure the amplitude of output signal keeping input signal amplitude constant for each frequency. Note down the output voltage for corresponding input frequency. Draw the graph between output voltage against frequency. As per Eq. (E.5) it is observed that the amplitude of the output signal decreases with increasing frequency. Note that as far as possible non-leaky capacitors made up of tantalum should be used in the feedback.

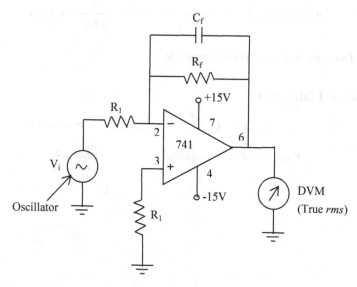

Fig. E.5 Practical integrator circuit using IC 741 Op-Amp

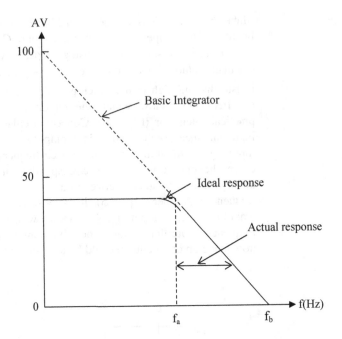

Fig. E.6 Frequency response of practical integrator

Observation Table (E.4)

$$R_1 = \underline{\hspace{1.5cm}} K, \quad C = \underline{\hspace{2cm}} uF, \quad V_i = \text{constant}$$

Sr. No.	Frequency of input signal (Hz)	Output voltage (V_0) (V)
1		
2		
3		
4		
5		
6		
-		
-		
-		
-		

Results and Discussion

From the graph of output voltage against frequency it is observed that the output voltage decreases with increasing frequency. The square waveform is applied at the input, and output is observed as in figure (Trace figure from CRO screen), which is ramp up and down, i.e. it generates the triangular wave. This shows that the circuit works as an integrator.

Precautions

1. Dual power supply must be regulated.
2. Apply very small input voltages so that the Op-Amp should not be saturated, because of the restrictions of amplifier that the amplified output should not go beyond supply voltage.
3. Measure all voltages with respect to ground.
4. Use high input impedance voltmeter (either analog or digital) for voltage measurement to avoid the loading.
5. Use proper values of R_f and R_1.
6. For applying the small voltages at the input terminal, a voltage divider network can be used.
7. Use non-leaky capacitors in the feedback circuit.
8. Selection of proper values of R_f, R_1 and C_1 is necessary to satisfy

$T \leq R_1 C_1$, time period of the input cycle.

Viva Questions

1. How integrator circuit works?
2. Explain action of capacitor connected in feedback in integrator circuit.
3. What will be the output of integrator when sinusoidal wave is applied at input?
4. What are the characteristics of ideal operational amplifier?
5. Why capacitor is shunted by resistor in feedback?
6. Define slew rate?
7. Whether slew rate plays any role in integration? Explain.

Experiment No. 5

Application of Op-Amp 741 as differentiator.

Aim   (i) To study Op-Amp as differentiator
      (ii) Observe the change in output voltage when variable frequency signal is applied at the input.

Apparatus Breadboard, IC 741 Op-Amp, resistors, capacitors, power supply ±15 V, oscillator, cathode ray oscilloscope (CRO) , hookup wire, etc.

Formula

$$\text{Output voltage} \quad V_0 = -R_f C_1 \frac{dV_i}{dt} \tag{E.7}$$

where V_i is the input voltage, R_f is the resistance connected in feedback, and C_1 is the capacitor connected in input. The negative sign indicates phase change of 180° between input and output signal. If a sinusoidal signal is applied at the input of the differentiator say

$$V_i = V \sin wt \tag{E.8}$$

The output will be

$$V_0 = -R_f C_1 V \omega \cos \omega t \qquad (E.9)$$

or

$$R_1 = \underline{\hspace{1cm}} K, C_1 = \underline{\hspace{1cm}} mF, R_f = \underline{\hspace{1cm}} K, C_f$$
$$= \underline{\hspace{1cm}} mF \qquad (E.10)$$

Equation (E.10) shows that the amplitude of the output signal of the differentiator is directly proportional to the frequency (ω) of the input signal. At low frequencies the capacitive reactance is high resulting in low gain or low output of the differentiator, whereas at high frequencies the output voltage increases with frequencies of input signal.

Procedure   Make the connections as shown in Figs. (E.7 and E.8). Select proper values of feedback resistor and capacitor as per the frequency response of the differentiator so that time period of the input signal $T \geq R_f C_1$ is satisfied. Calculate values of R_f and C_1 from $f_a = \frac{1}{2\pi R_f C_1}$ and $f_b = \frac{1}{2\pi R_1 C_1}$.

$f_b = 20 f_a$ and $C_1 < 1$ μF. Satisfying $R_1 C_1 = R_f C_f$ evaluate values of R_1 and C_f. f_a is the frequency at which gain of the differentiator is 0 dB and f_b is the gain limiting frequency.

The components R_1 and C_f are used for stability of the circuit and high-frequency noise correction. By selecting value of f_a as highest frequency to be differentiated and assuming $C_1 > 1$ μF, calculate value of R_1 and C_f.

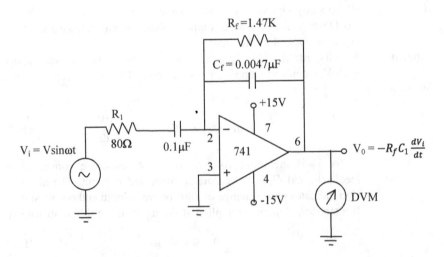

Fig. E.7 Practical differentiator using IC741 Op-Amp

Fig. E.8 Frequency response of the differentiation

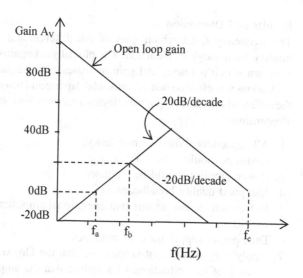

Connect these components in circuit, and switch on the power supply. Apply sinusoidal input to the differentiator through oscillator or function generator. Keeping amplitude of the input sin wave constant, change the frequency, and note the corresponding output voltage. Record the readings in observation Table (E.5), and plot the graph between voltage gains against frequency. As per Eq. (E.10) it is observed that the output voltage increases with increasing frequency of the input signal.

Observation Table (E.5)

$$R_1 = \text{_____}K, C_1 = \text{\_\_\_\_\_}uF, R_f = \text{_____}K, C_f = \text{\_\_\_\_\_}uF$$

Sr. No.	Frequency of input signal (Hz)	Output voltage (V_0) (V)
1		
2		
3		
4		
5		
6		
7		
-		
-		
-		
-		

Results and Discussion

The frequency f_a (at which gain of the differentiator is 0 dB) and f_b (the gain limiting frequency) is calculated. The plot of gain against frequency of input signal is drawn which is linear, and gain increases with frequency of input signal.

Various waveforms such as sinusoidal, square and triangular are feed at the input of the differentiator, and output waveforms such as cosine, spikes and square are drawn.

Precautions

1. All capacitors should be non-leaky.
2. Resistors should have 1% tolerance.
3. Power supplies should be regulated.
4. Use good quality breadboard.
5. Before connections ensure that operational amplifier IC should be in a working condition.
6. Dual power supply must be regulated.
7. Apply very small input voltages so that the Op-Amp should not be saturated, because of the restrictions of amplifier that the amplified output should not go beyond supply voltage.
8. Measure all voltages with respect to ground.
9. Use high input impedance voltmeter (either analog or digital) for voltage measurement to avoid the loading.
10. Use proper values of R_f and R_1.
11. For applying the small voltages at the input terminal, a voltage divider network can be used.

Viva Questions

1. Explain working of differentiator.

2. $T \geq R_f C_1$ should be satisfied for working of differentiator, explain.

3. What will be the output of differentiator for sinusoidal waveform applied at input?

4. What are the different parameters of operational amplifier?

5. What is CMRR?

Experiment No. 6

To design and study active filters using Op-Amp

Aim   (i) First-order low pass Butterworth filter
          (ii) Second-order low pass Butterworth filter
          (iii) First-order high pass Butterworth filter
          (iv) Second-order high pass Butterworth filter

Apparatus Breadboard, IC 741 Op-Amp, resistors, Capacitors, Power supply ± 15 V, oscillator or function generator, CRO, RMS DMM, hookup wire, etc.

Formula The higher cut-off frequency

$$f_H = \frac{1}{2\pi RC} \quad \text{first order low pass}$$

$$f_H = \frac{1}{2\pi\sqrt{R_1 R_3 C_1 C_2}} \quad \text{Second order low pass}$$

Lower cut-off frequency

$$f_L = \frac{1}{2\pi RC} \quad \text{first order high pass}$$

$$f_L = \frac{1}{2\pi\sqrt{R_1 R_3 C_1 C_2}} \quad \text{Second order high pass}$$

Procedure Initially design the filter circuit, and calculate values of resistor R and capacitor C. Select the higher cut-off frequency f_H, value of capacitor C, and calculate the value of resistor R from the given formula in first-order low pass Butterworth filter.

Similarly select the cut-off frequency f_H for second-order low pass filter, select the proper values of C_1 and C_2 capacitors, and calculate the values of R_2 and R_3 from the given formula. Preferably choose the value of capacitor C less than 1 µF.

The gain of the amplifier can be adjusted by proper selection of feedback resistor R_f and R_1 as per our desired value of voltage gain $\left(1 + \frac{R_t}{R_1}\right)$.

Make the connections as shown in Figs. E.9, E.10, E.11, E.12, E.13, E.14, E.15 and E.16 on breadboard, and apply ± 15V power supply to IC 741. Connect oscillator or function generator to the input and *RMS* DMM at the output. Change the frequency of the oscillator, and measure the output voltage at corresponding frequency. Record the observations in observation table.

Low pass filters
High pass filters (a) First order high pass Butterworth filter.
(b) Second-order high pass Butterworth filter.

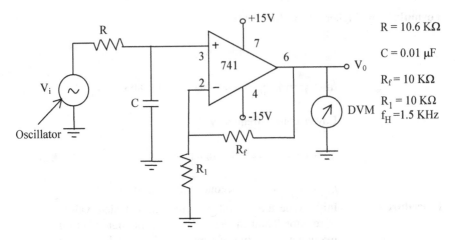

Fig. E.9 First-order low pass Butterworth filter

Observation Table E.6 (Low Pass/High Pass filters) Note: Use separate observation table for each filter circuit

$$V_i = _____V$$

Sr. No.	Frequency (Hz)	Output voltage V_0 (V)	Voltage gain in dB	Cut-off frequency Hz from formula
1				
2				
3				
4				
5				
6				
7				
8				
9				

Results and Discussion

The graph is plotted between voltage gains against frequency of the input signal, and cut-off frequency is calculated from the graph and formula. Both are found to be identical.

Precautions

1. Use high input impedance meter at the output.
2. Use good quality breadboard.
3. All connections should be clean and tight.

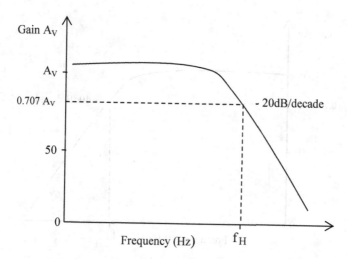

Fig. E.10 Frequency response of first-order low pass Butterworth filter

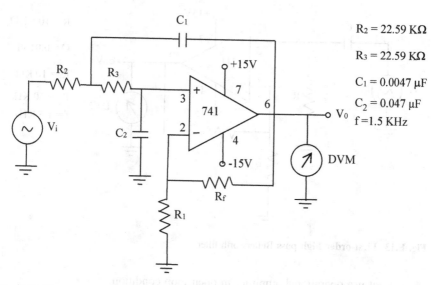

Fig. E.11 Second-order low pass Butterworth filter

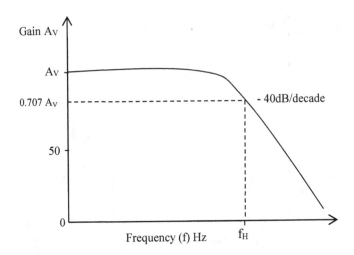

Fig. E.12 Frequency response of second-order low pass Butterworth filter

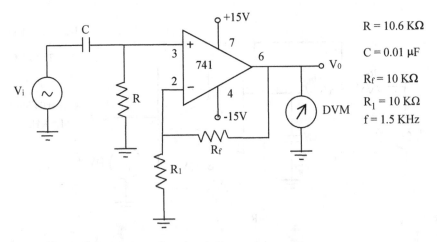

Fig. E.13 First-order high pass Butterworth filter

4. Do not use operational amplifier in open loop condition.
5. Dual power supply must be regulated.
6. Apply very small input voltage so that the Op-Amp should not be saturated, because of the restrictions of amplifier that the amplified output should not go beyond supply voltage.
7. Measure all voltages with respect to ground.

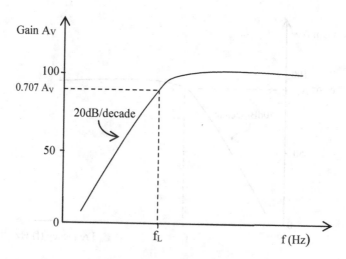

Fig. E.14 Frequency response of first-order high pass Butterworth filter

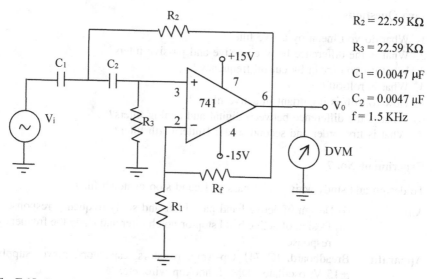

Fig. E.15 Second-order high pass Butterworth filter

8. Use high input impedance voltmeter (either analog or digital) for voltage measurement to avoid the loading.
9. Use proper values of R_f and R_1.
10. For applying the small voltage at the input terminal, a voltage divider network can be used.

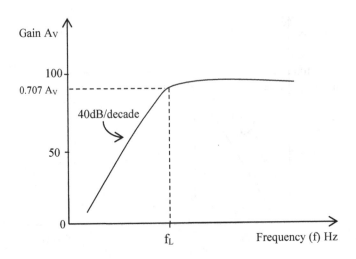

Fig. E.16 Second-order high pass Butterworth filter frequency response

Viva Questions

1. What do you mean by active filter?
2. What is the difference between active and passive filters?
3. What do you mean by cut-off frequency?
4. What is roll-off?
5. How roll-off is important in active filters?
6. What is the difference between digital and analog filters?
7. What is first-order and second- order Butterworth filter?

Experiment No. 7

To design and study active band pass and band stop or notch filter

Aim   (i) Design of active band pass filter and study frequency response
              (ii) Design of active band stop or notch filter and study the frequency
                  response

Apparatus   Breadboard, IC 741 Op-Amp, resistors, capacitors, power supply
              ±15 V, oscillator, DMM, hookup wire, etc.

Formula   (a) Band pass filter

$$\text{Centre frequency} \quad f_C = \sqrt{f_H f_L} \qquad (E.11)$$

$$\text{Quality factor} \quad Q = \frac{f_C}{f_H f_L} \qquad (E.12)$$

where f_c is called centre frequency, f_H is the higher cut-off, and f_L is
the lower cut-off frequency.

$$\text{Band width (BW)} = (f_H - f_L) \qquad \text{(E.13)}$$

(b) Band stop or Band reject or notch filter.
The same formula for centre frequency and bandwidth is used for wide band stop filter, but for narrow band reject filter where twin-T network is used as in Fig. E.20, the frequency at which maximum attenuation occurs is called notch frequency f_N, which is given by

$$f_N = \frac{1}{2\pi RC} \qquad \text{(E.14)}$$

Figures: (a) for band pass filter
By cascading high pass and low pass filters together a band pass filter can be formed.
(b) Band stop or notch filter.

Procedure   Make the connection as shown in Figs. E.17, E.18, E.19 and E.20 for respective filters on breadboard. In case of band pass filter for obtaining ±20 dB band pass, cascade first-order high pass and first order low pass filters. To obtain ±40 dB band pass cascade second-order high pass and second order low pass filters in series. Design filter circuit with proper component values of R and C.

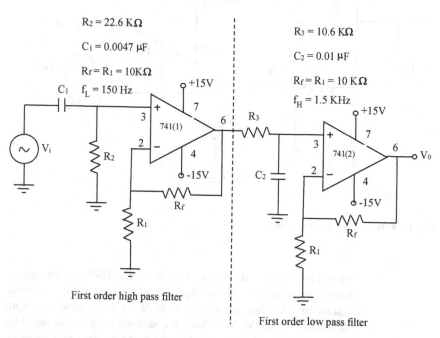

Fig. E.17 Band pass filter using 741 IC

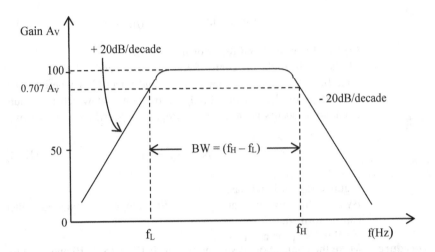

Fig. E.18 Frequency response of first order band pass filter

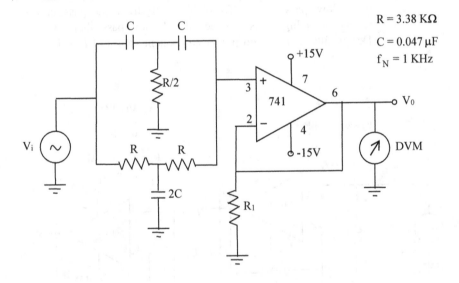

Fig. E.19 Band stop or notch T-filter

Connect oscillator at the input of the filter and DMM at the output. Change the frequency of the input signal keeping input amplitude constant for each frequency, and measure the corresponding output with the help of DMM. Note down the readings in observations table. Calculate the gain of the circuit. Plot the graph between voltage gains

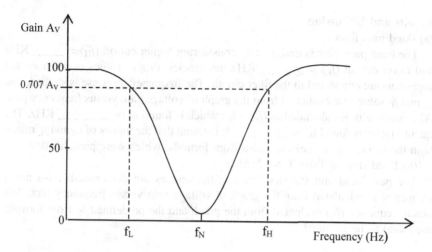

Fig. E.20 Frequency response of Twin-T notch filter

A_V against frequency f of the input signal. Determine the values of higher (f_H) and lower (f_L) cut-off frequencies. Calculate bandwidth and quality factor. Verify the experimental values of cut-off frequencies with the theoretical values initially decided while designing the filter circuit.

Observation Table

$R = $ _____K, $C = $ _____uF, $V_i = $ _____ V

$f_H = $ _____KHz, $f_L = $ _____KHz, $BW = $ _____KHz

Sr. No.	Frequency of input signal (Hz)	Output voltage V_0 (V)	Voltage gain A_V (dB)
1			
2			
3			
4			
5			
6			
7			

Results and Discussion

(a) Band pass filter

The band pass filter is designed by considering higher cut-off (f_H) = _____ KHz and lower cut-off (f_L) = _____ KHz frequencies. Proper values of resistors and capacitors are connected in the filter circuit. The frequency response is studied, and f_H and f_L values are evaluated from the graph of voltage gain versus frequency plot. Also bandwidth is calculated as $(f_H - f_L)$ which is found to be _____ KHz. The quality factor is found to be _____. It is found that the values of f_H and f_L match with the theoretical values evaluated from formula, which were predefined.

(b) Band stop or Twin-T notch filter

The pass band and the stop band of frequencies are determined. Also notch frequency is calculated from the graph of voltage gain versus frequency plot. The notch frequency (f_N) evaluated from the graph and the predefined f_N from formula are found to be identical.

Precautions

1. Dual power supply must be regulated.
2. Apply very small input voltage so that the Op-Amp should not be saturated, because of the restrictions of amplifier that the amplified output should not go beyond supply voltage.
3. Measure all voltages with respect to ground.
4. Use high input impedance voltmeter (either analog or digital) for voltage measurement to avoid the loading.
5. Use proper values of R_f and R_1.
6. For applying the small voltage at the input terminal, a voltage divider network can be used.

Viva Questions

1. Define band pass filter and notch filter.
2. Explain working of band pass and notch filter.
3. What is the difference between twin-T and Bridge-T filter?
4. What are the applications of band pass and notch filters?
5. In what way these filters are playing important role in bass and treble circuits of audio system?

Experiment No. 8

Study of Op-Amp as voltage to current converter and current to voltage converter

Aim   (i) Voltage to current converter

(ii) Current to voltage converter

Apparatus   Breadboard, IC 741, resistors, voltage source, milli-ammeter, hookup wire, power supply ± 15 V, etc.

Formula (i) Voltage to current converter

$$I_{out} = \frac{V_i}{R}$$ (E.15)

(ii) Current to voltage converter

$$V_0 = I_i R$$ (E.16)

Procedure Make the connections as shown in Figs. E.21 and E.22. Switch on the power supply. Vary input in the circuits. Note down the input voltage and corresponding output current in case of voltage to current converter, and note down the current (I) and the output voltage in

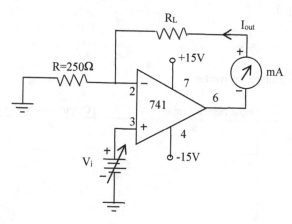

Fig. E.21 Voltage to current converter using Op-Amp

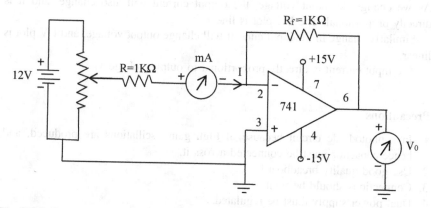

Fig. E.22 Current to voltage converter using IC 741 Op-Amp

case of current to voltage converter. Record the observations in observation table. Plot the graph between voltage and current in case of V to I converter and current versus voltage in case of I to V converter.

Observation Table

(a) Voltage to current converter

Sr. No.	Input voltage V_i (V)	Output current I (mA)
1		
2		
3		
4		
5		
6		

(b) Current to voltage converter

Sr. No.	Input current I (mA)	Output voltage V_0 (V)
1		
2		
3		
4		
5		
6		

Results and Discussion

As we change the input voltage, the output current will also change, and it is directly proportional. The V-I plot is linear.

Similarly change in the input current will change output voltage, and I-V plot is linear.

The input current is directly proportional to output voltage V_0.

Precautions

1. In a photodiode circuit because of high gain oscillations are produced, and hence capacitor must be connected across it.
2. Use good quality breadboard.
3. Connections should be tight.
4. Dual power supply must be regulated.

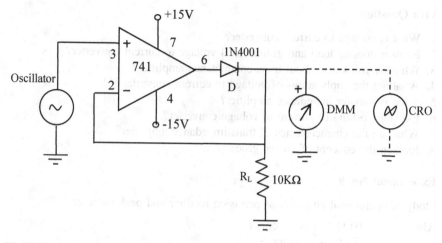

Fig. E.23 Precision rectifier or small signal half wave rectifier

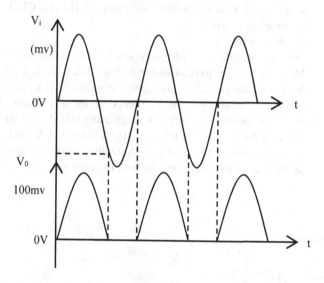

Fig. E.24 Input and corresponding output waveforms

5. Apply very small input voltages so that the Op-Amp should not be saturated, because of the restrictions of amplifier that the amplified output should not go beyond supply voltage.
6. Measure all voltages with respect to ground.
7. Use high input impedance voltmeter (either analog or digital) for voltage measurement to avoid the loading.
8. Use proper values of R_f and R_1.
9. For applying the small voltages at the input terminal, a voltage divider network can be used.

Viva Questions

1. What is voltage to current converter?
2. Explain floating load and ground load voltage to current converter?
3. What do you know about trans-conductance amplifier?
4. What are the applications of voltage to current converter V?
5. What is transfer impedance amplifier?
6. Explain working of current to voltage converter?
7. What are the characteristics of trans-impedance amplifier?
8. Explain the concept of virtual ground.

Experiment No. 9

Study of operational amplifier as precision rectifier and peak detector

Aim	(i) Half wave rectifier
	(ii) Peak detector
Apparatus	Breadboard, IC 741, resistors 1 KΩ, 10 KΩ, diode 1N41, power supply ±15 V, hookup wire, oscillator, DMM and CRO.
Figure	(a) Precision rectifier
	(b) Peak detector
Procedure	Precision rectifiers rectify very small AC voltage using Op-Amp. Make the connections as shown in Figs. E.23, E.24, E.25 and E.26. Switch on the power supply, apply very small AC signal say 20 mV of 1 KHz frequency to the rectifiers/detector, and observe the output on CRO. Measure the output voltage using DMM. Record the output voltages for 30, 40, 50, 60……. 100 mV up to 1 V, and note down corresponding output in observation table. Trace the input waveform as well as output waveform in both experiments.

(b) Peak detector –

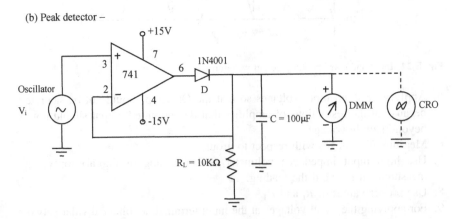

Fig. E.25 Peak detector using Op-Amp

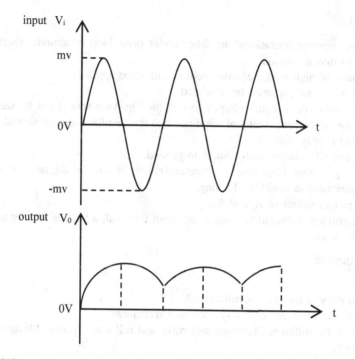

Fig. E.26 Input and output waveforms

Observation Table: (Half wave rectifier/precision rectifier) $f = 1$ kHz.

Sr. No.	Input signal strength (mV)	Output voltage V_0 (V)
1	20	
2	30	
3	40	
4	50	
5	60	
6	70	
–	–	
–	–	
–	–	
–	100 mV	

Results and Discussion

In ordinary rectifier sufficient input voltage is required to conduct the diode. Small magnitude AC signals cannot be rectified by ordinary diode alone. But if it is used in active rectifier circuits then very small AC signals can be rectified or detected.

Precautions

1. Do not operate operational amplifier under open loop condition; otherwise it will go into saturation.
2. Because of high gain, negative feedback must be applied.
3. Dual power supply must be regulated.
4. Apply very small input voltages so that the Op-Amp should not be saturated, because of the restrictions of amplifier that the amplified output should not go beyond supply voltage.
5. Measure all voltages with respect to ground.
6. Use high input impedance voltmeter (either analog or digital) for voltage measurement to avoid the loading.
7. Use proper values of R_f and R_1.
8. For applying the small voltage at the input terminal, a voltage divider network can be used.

Viva Questions

1. What do you mean by rectification?
2. Which elements are generally used as a rectifier?
3. What is the difference between half wave and full wave rectifier? Which one is better?
4. What is ripple factor?
5. Explain action of precision rectifier?
6. What is peak detector?
7. What are applications of peak detector?
8. Explain action of zero crossing detector?

Experiment No. 10

Application of Op-Amp as instrumentation amplifier for measurement of temperature.

Aim   To measure the temperature of a bath.

Apparatus   Breadboard, IC Op-Amp 741, resistors, thermistor, DC power supply ±15 V, DMM, test tube, hookup wire, toluene 30 ml, beaker, sand bath, heater, dimmerstat, connecting wires, thermometer, etc.

Formula

$$\text{output voltage} \quad V_0 = -\frac{R_f}{R_1}(V_a - V_b)$$

Procedure   Make the connections on the breadboard as shown in Fig. E.27. Wheatstone bridge is used in the circuit where in one arm thermistor R_x is connected. All resistors in the branch have nearly same value. Digital multimeter is connected at the output to measure the

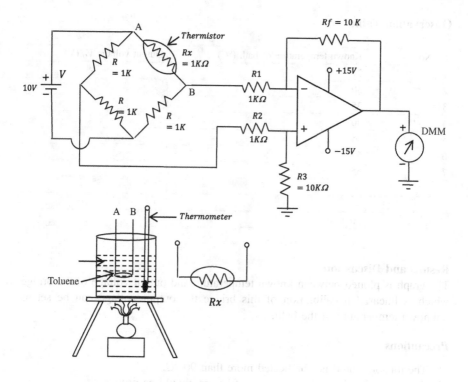

Fig. E.27 Bridge instrumentation amplifier

unbalanced output voltage. The Wheatstone bridge is excited by externally connected 10 V DC voltage. ±15 V power supply is used for Op-Amp.

To measure the temperature of a toluene bath, take 30 ml toluene in a big test tube. Dip thermistor in toluene bath along with thermometer to know the temperature of bath. Take water in a 250 ml beaker, and keep this toluene filled test tube in beaker in which thermistor is placed along with thermometer. Heat the water so that thermistor will be heated. Initially note down the output voltage for room temperature. As soon as thermistor is heated the bridge will be unbalanced, and more voltage will be shown by DMM.

Record the output voltage with known temperature of toluene. Plot the graph between known temperature against output voltage which will be linear. Note down the readings at the interval of 5 °C in observation table.

Observation Table

Sr. No.	Known temperature of bath (°C)	Output voltage V_0 (V)
1	27	
2	30	
3	35	
4	40	
5	45	
6	50	
7	55	
-	--	
-	--	
-	80	

Results and Discussion

The graph is plotted between known temperature and output voltage of the bridge which is linear. On calibration of this bridge the output voltage can be set as unknown temperature of the bath.

Precautions

1. Thermistor should not be heated more than 90 °C.
2. Selection of resistors in four arms of bridge should be proper.
3. Adjust proper gain of the operational amplifier.
4. Digital multimeter should have large input impedance.
5. Dual power supply must be regulated.
6. Apply very small input voltage so that the Op-Amp should not be saturated, because of the restrictions of amplifier that the amplified output should not go beyond supply voltage.
7. Measure all voltages with respect to ground.
8. Use high input impedance voltmeter (either analog or digital) for voltage measurement to avoid the loading.
9. Use proper values of R_f and R_1.
10. For applying the small voltage at the input terminal, a voltage divider network can be used.

Viva Questions

1. Which material is used for fabrication of thermistor?
2. Why toluene is used as bath?
3. What do you mean by instrumentation amplifier?
4. What are the special features of instrumentation amplifier?
5. Will it be possible to replace DC source by AC?
6. What do you mean by calibration?
7. What do you mean by resolution of instrument?

Experiment No. 11

Design and study of operational amplifier as phase shift oscillator.

Aim   (i) Design phase shift oscillator

                (ii) Measure frequency of output signal.

Apparatus   Breadboard, IC 741, several resistors, capacitors, potentiometer 1 MΩ, power supply, CRO, hookup wire, etc.

Formula

$$\text{Voltage gain}\quad A_V = \left|\frac{R_f}{R_1}\right| = 29 \tag{E.17}$$

$$\text{Frequency of oscillation}\quad f_0 = \frac{1}{2\pi\sqrt{6}RC} \tag{E.18}$$

Figure

Procedure   Make the connections as shown in Fig. E.28. Select frequency of oscillator f_0 and value of capacitor C, and then calculate value of resistor R from the given formula $f_0 = \frac{1}{2\pi\sqrt{6}RC}$. Maintain the gain of the amplifier as $R_f = 29R_1$ by selection. Generally 1 MΩ potentiometer is used (wire wound is preferable). Connect cathode ray oscilloscope (CRO) at the output to observe the waveform. Switch on the power supply and adjust the voltage gain of the amplifier to 29 by adjusting the potentiometer resistance to $29R_1$, so that oscillation can start.

Calculate the frequency of oscillation from CRO either by measuring the time period of the cycle or by lissajous figure. Note down the

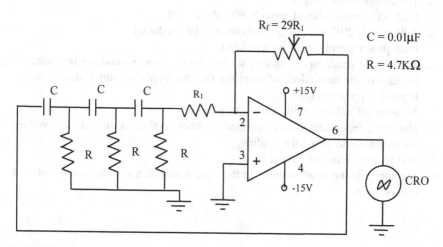

Fig. E.28 Phase shift oscillator

reading in the observation table. Verify the experimental frequency with the selected frequency (f_0) of oscillation.

Repeat this experiment for various values of R or C combination, and verify the frequency of oscillation.

Observation table

Sr. No.	Value of resistor (KΩ)	Value of capacitor (µF)	Experimental frequency observed on CRO (KHz)	Frequency selected from formula f_0 (KHz)
1				
2				
3				
4				
5				

Results and Discussion

The phase shift oscillator is designed, and waveform is observed on CRO which is sinusoidal. The frequency is calculated from CRO by measuring the time period of the cycle. The value of frequency of oscillation f_0 is verified with the theoretical value selected initially. It is found that both frequencies are identical.

Precautions

1. As feedback is small, starting oscillations is difficult.
2. The output is also small.
3. Each RC section should provide 60° phase shift.
4. Adjust $R_f = 29R_1$ so that oscillations will be produced.
5. Dual power supply must be regulated.
6. Apply very small input voltages so that the Op-Amp should not be saturated, because of the restrictions of amplifier that the amplified output should not go beyond supply voltage.
7. Measure all voltages with respect to ground.
8. Use high input impedance voltmeter (either analog or digital) for voltage measurement to avoid the loading.
9. Use proper values of R_f and R_1.
10. For applying the small voltages at the input terminal, a voltage divider network can be used.

Viva Questions

1. What is feedback?
2. What are the different types of feedback?
3. Which feedback is used in oscillator?
4. Why above circuit is called phase shift oscillator?
5. What is a Barkhausen criterion?
6. Why it is necessary to adjust the gain of the amplifier as 29?
7. What is lissajous figure? How it is formed?
8. On which factors the frequency of oscillation depends?
9. How much phase shift is added by each RC network?
10. Why 180° phase shift is required?

Experiment No. 12

To design and construct the Wien bridge oscillator using Op-Amp.

Aim (i) To design Wien bridge oscillator

(ii) To measure the frequency of output signal

Apparatus Breadboard, IC 741, resistors, capacitors, 1 KΩ potentiometer, power supply ±15 V, CRO, hookup wire, etc.

Formula Voltage gain of the non-inverting amplifier is given by (Fig. E.29)

$$A_V = 1 + \frac{R_f}{R_1} = 3 \qquad (E.19)$$

Or

$$R_f = 2R_1 \qquad (E.20)$$

Frequency of oscillation f_0 is

$$f_0 = \frac{1}{2\pi RC} \qquad (E.21)$$

Procedure and Observation Table

Experimental procedure is same as given in phase shift oscillator experiment no. 11. The observation table is also similar to phase shift oscillator experiment.

Results and Discussion

This part is also similar to phase shift oscillator.

The procedure, observations and results of Wien bridge oscillator are similar to that of phase shift oscillator experiment.

Fig. E.29 Wien bridge
oscillator using Op-Amp

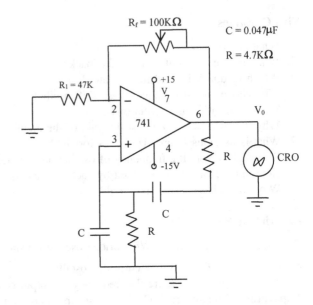

Precautions

1. Wien bridge oscillator offers zero phase shift, and hence non-inverting mode of Op-Amp is preferred.
2. While resonating the voltages at non-inverting and inverting terminals should be equal and in phase with each other.
3. Gain should be greater than 3 to start the oscillation.
4. There is a frequency limit of 1 MHz.
5. Dual power supply must be regulated.
6. Apply very small input voltage so that the Op-Amp should not be saturated, because of the restrictions of amplifier that the amplified output should not go beyond supply voltage.
7. Measure all voltages with respect to ground.
8. Use high input impedance voltmeter (either analog or digital) for voltage measurement to avoid the loading.
9. Use proper values of R_f and R_1.

Viva Questions

1. Why gain is adjusted to 3?
2. On which factors the frequency of oscillations depends?
3. What will be the frequency range of Wien bridge oscillator?
4. Explain the working of the oscillator.

Experiment No. 13

Design and study of astable multivibrator using Op-Amp (Square wave generator)

Aim   (i) Design of astable multivibrator

(ii) Measure on and off time of the square wave.

Apparatus Breadboard, IC 741 Op-Amp, resistors, capacitors, CRO, hookup
wire, power supply ±15 V, etc.

Formula

$$\text{Frequency of oscillator} \quad f = \frac{1}{2RC} = \frac{1}{T} \qquad (E.22)$$

Where T is time period of one cycle.

$$T = T_{ON} + T_{OFF} \qquad (E.23)$$

Figure

Procedure The astable multivibrator is called free-running multivibrator or
square wave generator. The resistor R and capacitor C decide the
frequency of oscillation. For proper oscillations the potentiometer of
22 KΩ is used. By adjusting the value of resistor the proper feedback
is provided.

Make the connections as shown in Fig. E.30. Switch on the power
supply, and adjust the resistance R_1 to get the proper feedback so that
oscillations will be produced at the output. Connect cathode ray
oscilloscope (CRO) at the output, and observe the square wave

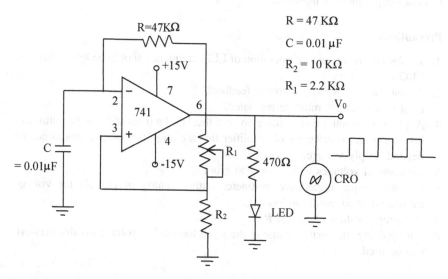

Fig. E.30 Square wave generator using Op-Amp

generated. Measure *on* and *off* time of the pulse, and calculate the frequency of oscillation. If the frequency of oscillation is less having *on* and *off* time 3–4 *s* or more, then to observe the output, a LED in series with 470 Ω resistor can be connected. In this case stopwatch is necessary to measure *on* and *off* time.

Record the observations in observation table. Change *R* and *C* values or any one component will change the frequency of oscillation. Perform this experiment for various values of *R* and *C*, and record the observation for calculation of frequency.

Observation Table

Sr. No.	Value of resistor R (KΩ)	Value of capacitor C (µF)	On time T_{ON} (s)	Off time T_{OFF} (s)	Time period T (s)	Frequency (Hz)
1						
2						
3						
4						
5						

Results and Discussion
The square wave is observed on CRO, and *on* and *off* time are measured. Frequency of oscillation of the waveform is found to be _____ Hz. By changing the *R* and *C* values experiment is repeated.

Precautions

1. To observe the on and off operation of LED, frequency should be low; otherwise CRO is necessary.
2. Adjust value of R_1 to get proper feedback.
3. Dual power supply must be regulated.
4. Apply very small input voltage so that the Op-Amp should not be saturated, because of the restrictions of amplifier that the amplified output should not go beyond supply voltage.
5. Measure all voltages with respect to ground.
6. Use high input impedance voltmeter (either analog or digital) for voltage measurement to avoid the loading.
7. Use proper values of R_f and R_1.
8. For applying the small voltage at the input terminal, a voltage divider network can be used.

Viva Questions

1. What are non-sinusoidal oscillators?
2. How square wave is generated?
3. What do you mean by symmetrical and unsymmetrical multivibrators?
4. Describe working of astable multivibrator in brief.

Experiment No. 14

Design and study of triangular wave generator using Op-Amp

Aim	(i) Design of triangular waveform generator
	(ii) Measure the frequency of the oscillations.
Apparatus	Breadboard, IC 741 Op-Amp, resistors, capacitors, power supply ±15 V, CRO, hookup wires, etc.
Formula	frequency of oscillations

$$f = \frac{R_3}{4R_1 C_1 R_2}$$

Figure

Procedure   Make the connections as shown in Fig. E.31. Connect proper values of R_1, R_2, R_3 and C_1 in the circuit. Switch on the power supply. Connect CRO at the output to observe the waveform. The first Op-Amp will act as comparator, i.e. zero crossing detector because inverting terminal of the Op-Amp is connected to ground. Second Op-Amp acts as integrator which converts square wave into triangular waveform. Adjust the resistance of the potentiometer R_3 to get square wave output at the first Op-Amp. The frequency of oscillations can be adjusted by changing the value of R_3.

Record the observations for various values of R_3 in observation table, and calculate the frequency of oscillations f.

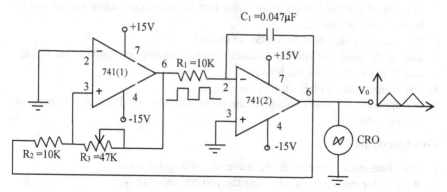

Fig. E.31 Triangular waveform generator

NOTE: If the *on* time of the square wave is adjusted to be small, the oscillator will work as saw tooth waveform generator. Rest of the procedure and experiment is same as triangular waveform generator.

Observation Table

Sr. No.	Resistor values (KΩ)			Value of capacitor C (µF)	Frequency f (Hz) from formula	Frequency measured on CRO (Hz)
	R_1	R_2	R_3			
1						
2						
3						
4						
5						
6						

Results and Discussion

By changing the values of R_1, R_2 and R_3 frequency of oscillation is changed. The frequency calculated from formula and measured frequency on CRO are found to be identical.

Precautions

1. Measure accurate time period of cycle on CRO.
2. Select proper values of resistors and capacitors.
3. To start the oscillations adjust value of R_3.
4. Dual power supply must be regulated.
5. Apply very small input voltage so that the Op-Amp should not be saturated, because of the restrictions of amplifier that the amplified output should not go beyond supply voltage.
6. Measure all voltages with respect to ground.
7. Use high input impedance voltmeter (either analog or digital) for voltage measurement to avoid the loading.
8. Use proper values of R_f and R_1.
9. For applying the small voltage at the input terminal, a voltage divider network can be used.

Viva Questions

1. How integrator converts square wave into triangular wave?
2. What is the necessity of adjusting the resistor value of R_3?
3. Explain the working of comparator.
4. What is duty cycle?

5. What will be the effect of change in duty cycle on output waveform?
6. Explain working of integrator.
7. How ramp up and ramp down voltage is generated?
8. List the factors which decide the frequency of oscillation of triangular wave?

Experiment No. 15

Study of Schmitt trigger using Op-Amp.

Aim   (i) Determination of lower trip (LTP) and
         (ii) Upper trip (UTP) points

Apparatus   Breadboard, IC741 Op-Amp, resistors, Power supply ±15 V,
                Rheostat 1 KΩ, or variable power supply, DMM, hookup wires, etc.

Formula   (i) Upper threshold voltage UTP

$$V_{\text{UTP}} = \left(\frac{R_1}{R_1 + R_2}\right) V_{\text{sat}}^+ \tag{E.24}$$

(ii) Lower threshold voltage LTP

$$V_{\text{LTP}} = \left(\frac{R_1}{R_1 + R_2}\right) V_{\text{sat}}^- \tag{E.25}$$

(iii) The hysteresis voltage

$$= V_{\text{UTP}} - V_{\text{LTP}} \tag{E.26}$$

Figure

Procedure
Make the connections as shown in Figs. E.32 and E.33. Switch on the power supply, and apply ±15 V DC voltage to the 10 KΩ rheostat. The variable terminal of the rheostat is connected to the inverting terminal of Op-Amp. Vary the input voltage in steps from negative to positive, and note down the output voltage (V_0) for every step of input voltage (V_i). Do not move the variable (voltage) terminal in reverse direction when you are recording the readings in forward direction and vice-versa. Write down the observations during forward mode. Then reverse the position of wiper of rheostat and note down readings in reverse mode. Note down the input voltage and corresponding output voltage in observation table. Draw the hysteresis, and determine Upper threshold or trip voltage (V_{UTP}) and lower threshold or trip voltage (V_{LTP}). Calculate hysteresis voltage from the graph.

   Apply sinusoidal signal at the input of Schmitt trigger, and observe the output waveform on CRO and trace it.

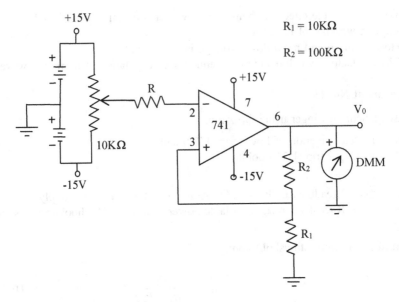

Fig. E.32 a Schmitt trigger using IC 741

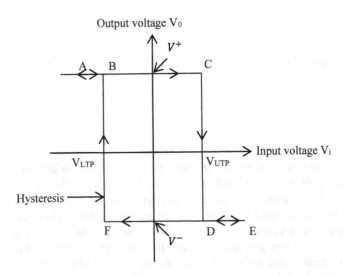

Fig. E.33 b Hysteresis plot

Observations Table

Sr No.	Input voltage V_i(V)	Output voltage V_o (V)
1		
2		
3		
4		

Results and Discussions

The graph between input voltage (V_i) and output voltage (V_0) is plotted. The upper threshold voltage (V_{UTP}) and lower threshold voltage (V_{LTP}) are determined, and hysteresis voltage is calculated.

The V_{UTP} is found to be _____ V

The V_{LTP} is found to be _____ V

The hysteresis voltage is found to be _____ V

Precautions

1. While performing experiment do not change the position of rheostat in backward direction while changing it in forward and vice versa.
2. Select proper values of R_1 and R_2 to get LTP and UTP.
3. Dual power supply must be regulated.
4. Apply very small input voltage so that the Op-Amp should not be saturated, because of the restrictions of amplifier that the amplified output should not go beyond supply voltage.
5. Measure all voltages with respect to ground.
6. Use high input impedance voltmeter (either analog or digital) for voltage measurement to avoid the loading.
7. Use proper values of R_f and R_1.
8. For applying the small voltage at the input terminal, a voltage divider network can be used.

Viva Questions

1. What do you mean by upper threshold voltage (UTP) and lower threshold voltage (LTP)?
2. What do you mean by hysteresis?
3. What will be the output when irregular wave is applied at the input?
4. Explain working of Schmitt trigger.
5. What are the applications of Schmitt trigger?
6. What do you mean by saturation voltage?

Experiment No. 16

To study voltage to frequency (V/F) converter using Op-Amp.

Aim   (i) Design and construct V/F converter
        (ii) Determine frequency conversion rate per volt.

Apparatus   Breadboard, IC 741 Op-Amp, IC 555 timer, power supply ± 15 and +6 V, resistors, capacitors, hookup wire, CRO, DVM, etc.

Figure

Procedure   The voltage to frequency converters are voltage-controlled oscillators (VCO). IC 9400 is the best IC by which pulse and square wave outputs up to 100 KHz frequency can be obtained. The output frequency can be controlled by the input voltage. This IC consists of several operational amplifiers. The simple practical circuit of V/F converter is shown in Fig. E.34. In this circuit single Op-Amp is used as comparator and IC 555 timer as one shot or monostable multivibrator.

Make the connections as shown in Fig. E.34. Switch on the power supply. Apply input voltage to non-inverting terminal of Op-Amp through resistor R_4 by variable power supply. Variations can be made from 0 to 5 V. Connect CRO at the output, measure the input voltage applied, and observe the square wave on the CRO. Note down the frequency of the square wave. Change the input voltage in steps, and note down the frequency on CRO. Perform this experiment up to 5 V. Record you observations in observation table. Plot the graph between frequency of output waveform and DC input voltage. This

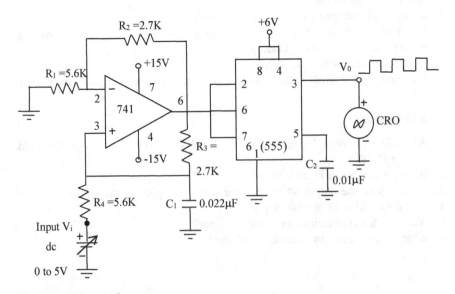

Fig. E.34 Voltage to frequency converter

plot is linear. Calculate the rate of change of frequency with input voltage.

Observation Table

Sr. No.	Input voltage V_i (V)	Frequency of output signal f (Hz)	Output amplitude V_0 (V)
1			
2			
3			
4			
5			
6			

Results and Discussion

The plot of frequency of output waveform with input voltage applied is found to be linear. The rate of change of frequency with input voltage is found to be _____ KHz/V.

Precautions

1. In ADVFC 32 IC voltage to frequency (V/F) and frequency to voltage (F/V) converter charge balancing circuits are used, and hence proper precautions must be taken because of electrostatic discharge sensitivity of the device. This may give electric shock of 4000 V to human body or test equipment.
2. Dual power supply must be regulated.
3. Apply very small input voltage so that the Op-Amp should not be saturated, because of the restrictions of amplifier that the amplified output should not go beyond supply voltage.
4. Measure all voltages with respect to ground.
5. Use high input impedance voltmeter (either analog or digital) for voltage measurement to avoid the loading.
6. Use proper values of R_f and R_1.
7. For applying the small voltage at the input terminal, a voltage divider network can be used.

Viva Questions

1. What is voltage-controlled oscillator?
2. What is the function of IC 555 timer?
3. Will it be possible to obtain sinusoidal wave at output?

4. Which factor decides the rate of change of frequency?
5. What do you know about the comparator?
6. Why monostable multivibrator is connected?

Experiment No. 17

To study the frequency to voltage (*F/V*) converter using Op-Amp

Aim	(i) Design and construct the *F/V* converter
	(ii) Calculate the change in output with input frequency change.
Apparatus	Breadboard, IC 741 Op-Amp, resistors, capacitors, oscillator, DVM, hookup wires, etc.
Figure	
Procedure	The first stage Op-Amp is working as Schmitt trigger and second stage as integrator. The sinusoidal wave is converted into square waveform by the Schmitt trigger, and integrator produces a ramp.

Make the connections as shown in Fig. E.35. The oscillator is connected at the input, and the sinusoidal signal is directly connected to the inverting terminal of the first stage Op-Amp. Change the frequency of the input signal, and adjust the amplitude to get square wave output of the Schmitt trigger. Note down the output voltage for corresponding input signal frequency. Record the observations in observation table. Plot the graph between input signal frequency and output voltage. A linear plot will give you change in output voltage per frequency of input signal.

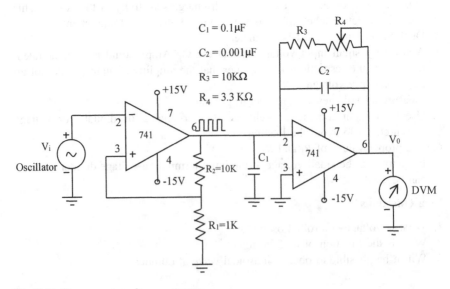

Fig. E.35 Frequency to voltage converter

Observation Table

$$V_i = \underline{\hspace{2cm}} V$$

Sr. No.	Frequency of input signal (Hz)	Output voltage V_0 (V)
1		
2		
3		
4		
5		
6		

Results and Discussion

The output voltage varies linearly with increasing frequency of the input signal. The change in output voltage with change in input frequency of the signal is found to be _____.

Precautions

1. In ADVFC 32 IC voltage to frequency (V/F) and frequency to voltage (F/ V) converter charge balancing circuits are used, and hence proper precautions must be taken because of electrostatic discharge sensitivity of the device. This may give electric shock of 40 V to human body for or test equipment.
2. Dual power supply must be regulated.
3. Apply very small input voltage so that the Op-Amp should not be saturated, because of the restrictions of amplifier that the amplified output should not go beyond supply voltage.
4. Measure all voltages with respect to ground.
5. Use high input impedance voltmeter (either analog or digital) for voltage measurement to avoid the loading.
6. Use proper values of R_f and R_1.
7. For applying the small voltage at the input terminal, a voltage divider network can be used.

Viva Questions

1. What is the function of integrator?
2. How IC 741 operational amplifier converts sinusoidal waveform into square waveform?
3. Are there any restrictions on integrator?
4. What is the resolution of the circuit?

Appendix II: Pin Configuration of ICs

IC 741

The IC 741 is an operational amplifier designed by fair child semiconductors. It is a versatile amplifier used as a multipurpose device in many electronic instruments/equipments. The pin configuration of this amplifier is given in Fig. A.36.

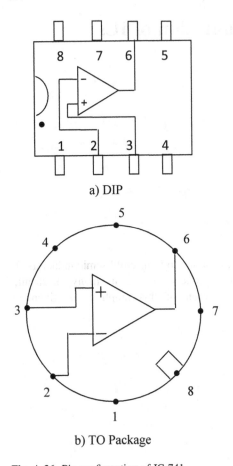

Pin No.		Function
1	-	Offset null
2	-	Inverting input
3	-	Non-inverting input
4	-	$-V_{CC}$
5	-	Offset null
6	-	Output
7	-	$+V_{CC}$
8	-	Not connected

a) DIP

b) TO Package

Fig. A.36 Pin configuration of IC 741

Appendix III

DC regulated power supply (**±15 V**)

Operational amplifier requires a ±15 V DC regulated power supply. Simple ±15 V regulated power supply can be constructed using very popular IC's of 78XX and 79XX series, where XX denotes the regulated voltage of IC. A fixed ±15 V DC regulated power supply which gives better regulation and has current capacity of 1 A can be designed and constructed in the laboratory with minimum number of components. Figure A.37 shows the circuit of ± 15V DC regulated power supply.

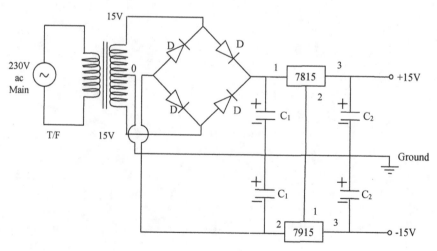

Fig. A.37 Dual power supply ±15 V

© The Editor(s) (if applicable) and The Author(s), under exclusive license to
Springer Nature Singapore Pte Ltd. 2022
S. Yawale and S. Yawale, *Operational Amplifier*,
https://doi.org/10.1007/978-981-16-4185-5

Fig. A.38 ICs 78XX and
79XX pin configuration

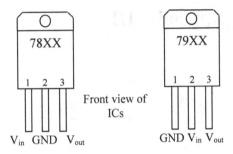

Front view of
ICs

List of Components

Transformer (T/F)	230 V ac—PP, 15 V-0-15 V—SS, 1 Amp
Diodes	1N4004 Or BY 127
IC's	LM7815 and LM7915
Capacitors	C_1 – 1000 μF/40 V, C_2 – 100 μF/50 V

IC LM7815 is a three terminal 1 A positive voltage regulator, and LM7915 is also a three terminal 1 A negative voltage regulator. These ICs give ±15 V regulated voltage.

The pin configuration of 78XX and 79XX ICs is given in Fig. A.38.

A simple printed circuit board can be used for the design and assembling of the power supply. These components are easily available in the market with low cost.

References

1. E.J. Kennedy, *Operational Amplifier Circuits; Theory and Applications* (Oxford University Press, Oxford, 1995)
2. H. Johan, *Operational Amplifiers; Theory and Design*, 2nd ed. (Springer, 2017)
3. D. Terrell, *Op-Amps: Design, Application and Troubleshooting Newness* (1996)
4. S. Antoch, *Op-Amp Circuits: Simulations and Experiments* Ed. ZAP Studio (2016)
5. M. Grabel, *Microelectronics*, 2nd edn. (McGraw Hill Book Co., New York, 1988)
6. P.R. Gray, R.G. Mayer, *Analysis and Design of Analog Integrated Circuits*, 2nd edn. (Wiley and Sons, New York, 1984)
7. A.S. Sedra, K.C. Smith, *Microelectronics Circuits, Holt* (Rinehart and Winston Inc., New York, 1981)
8. S. Soclof, *Analog Integrated Circuits* (Prentice-Hall, Englewood Cliffs, 1985)
9. J.K. Roberge, *Operational Amplifiers: Theory and Practice* (Wiley, New York, 1975)
10. J. Millman, H. Taub, *Pulse, Digital and Switching Waveforms* (McGraw Hill Book Co., New York, 1965)
11. J. Millman, C.C. Halkias, *Integrated Electronics: Analog and Digital Circuits and Systems* (McGraw Hill Book Co., New York, 1972)
12. R.A. Gayakwad, *Op-Amp and Linear Integrated Circuits*, 3rd edn. (Prentice Hall of India, New Delhi, 1988)
13. M. Kahn, *The Versatile Op-Amp. Holt* (Rinehart and Winston Inc., New York, 1970)
14. W.D. Stanley, *Operational Amplifiers with Integrated Circuits* (Merrill Pub. Co., Columbus, 1989)
15. C. George, S. Winder, *Operational Amplifiers*, 5th edn. (Newness, New York, 23)
16. A.S. Sedra, K.C. Smith, *Microelectronics Circuits*, 4th edn. (Oxford University Press, New York, 1998)
17. S. Rosenstark, *Feedback Amplifier Principles* (MacMillan, New York, 1986)
18. P.J. Hurst, Exact simulation of feedback circuit parameters. IEEE Tran. Circ. Syst. **38**, 1382–1389 (1991)
19. P.J. Hurst, A comparison of two approaches to feedback circuit analysis. IEEE Trans. Educ. **35**, 253–261 (1992)
20. J.L. Rodriguez Marrero, Simplified analysis of feedback amplifiers. IEEE Trans. Educ. **48**(1), 53–59 (25)
21. K. Nandini, Jog, *Electronics in Medicine and Biomedical Instrumentation* (Prentice Hall India, 26)
22. K.V.T. Piipponen, R. Sepponen, P. Eskelinen, IEEE Trans. Biomed. Eng. **54**(10), 1822–1828 (27)

23. Texas Instruments Application Report, Noise Analysis in Operational Amplifier Circuits, SLVA043A, 1999, SLOA082.
24. B. Carter, Texas Instruments Application Report, Noise Analysis in Operational Amplifier Circuits, SLOA082
25. D.F. Stout, *Handbook of Operational amplifier Design*, (McGraw-Hill, 1976), pp. 45–51
26. G.B. Clayton, B.W.G. Newby, B.H. Newnes, *Operational Amplifiers* (1992)
27. B.D.H. Tellegen, The gyrator. A new electric element. Philips Res. Rept. **3**(81–101), 388–393 (1948)
28. B.T. Morrison, *An Introduction to Gyrator Theory" Popular Electronics* (1977), pp. 58–59
29. T.H. Lynch, The Right Gyrator Trims the Fat off Active Filters' Electronics (1977), pp. 115–119
30. Application Report, Texas instruments, Noise analysis in operational amplifier circuits (27)
31. T.J. Sobering, *Op-Amp Noise Analysis*, Technical Note 5, May 1999
32. W.G. Jung, *Op-Amp Applications* (22), pp. 1.76–1.87
33. L. Smith, D.H. Sheingold, Analog dialogue **3**(1), 3–15 (1969)
34. J.G. Graeme, G.E. Tobey, L.P. Huelsman, *Operational Amplifier, Design and Applications* (Mc Grow Hill Int. Book Co. 1971), pp. 139–149
35. J.V. Vait, L.P. Huelsman, G.A Korn, *Introduction to Operational Amplifier, Theory and Applications* (Mc Grow Hill Kogakusha Ltd., Tokyo 1975), pp. 107–118
36. G. Giusi, F. Crupi, C. Pace, P. Magnone, IEEE Trans. Circ. Syst. **56**(1), 97–101 (29)
37. K.K. Asparuhova, E.A. Gadjeva, *27th International Spring Seminar on Electronics Technology* (24)
38. Texas Instruments Application Report, Op-Amp Noise Theory and Applications SLO A082
39. H. Johan, *Operational Amplifiers, Theory and Design*, 2nd edn. (Springer Science, 2011)
40. K. Plipponen, R. Sepponen, P. Eskelinen, IEEE Trans. Biomed. Eng. **54**(10), 1822–1825 (27)

Subject Index

Printed in the United States
by Baker & Taylor Publisher Services

Printed in the United States
by Baker & Taylor Publisher Services